PURPOSE IN A

Psychology

Editor
GEORGE WESTBY
Professor of Psychology
University College of South Wales, Cardiff

PURPOSE IN
ANIMAL BEHAVIOUR

F. V. Smith
Professor of Psychology
University of Durham

HUTCHINSON UNIVERSITY LIBRARY
LONDON

HUTCHINSON & CO (*Publishers*) **LTD**
3 Fitzroy Square, London W1

London Melbourne Sydney Auckland
Wellington Johannesburg Cape Town
and agencies throughout the world

First published 1971

*This book has been set in Times type, printed in Great Britain
on smooth wove paper by Anchor Press, and
bound by Wm. Brendon, both of Tiptree, Essex*

ISBN 0 09 109130 6 (cased)
0 09 109131 4 (paper)

CONTENTS

PREFACE

One of the chief problems in a work of this size is to decide what to omit. It is hoped nevertheless that sufficient information has been included to provide the reader with a vista of different species contending with similar basic problems but with different behavioural resources. Recurring throughout studies of behaviour is the issue of the degree to which purpose and awareness are involved. Unless they are ignored, important decisions as to methodology and explanation arise and the work will have in part succeeded if the reader acquires a greater awareness of these problems.

My thanks are due to Dr J. D. Delius, Dr Eric Sunderland and Dr J. C. Coulson for various points of advice, to Dr P. R. Evans for help with the proofs and especially to Dr Michael H. Day for his generous guidance on the illustrations of early man. They are in no sense responsible for any errors which the book may contain. I must also thank Mrs Barbara Atkinson for giving up her private time to type the bibliography and index, Mr Malcolm Rolling for photographic work with the illustrations and Miss Margaret Watt for the preparation of Fig. 1. Some mention of the tolerance of Miss Ann Douglas of Hutchinson University Library and of Professor George Westby is also appropriate.

Grateful acknowledgment is made to the authors Sir Julian Huxley, Dr F. R. Walther and Dr K. E. L. Simmons and to the following publishing houses for permission to use illustrations from their works:

E. J. Brill (*Behaviour*), University of Chicago Press, Harvard University Press, Oxford University Press, Academic Press, W. H.

Freeman & Co. (*Scientific American*), Paul Parey (*Zeitschrift für Tierpsychologie*), J. F. Bergmann—Verlagsbuchhandlung, Springer-Verlag.

Durham,   F.V.S.
July, 1971

I

INTRODUCTION

When one has solved a problem of any kind or is seeking an answer, the solutions of others contending with the same or comparable problems is usually a matter of genuine human interest. In the days before the opening of any international aircraft, motor, or boat show or indeed fashion show, public interest has frequently to be restrained and designers attend the opening not only to gauge the reception of their own productions but also to see how others have contended with kindred problems which include the stresses of use, probable demand, the cost of materials and the resources of skill and technology which can be devoted to such projects.

In a comparable way, students of the social sciences have a genuine interest in the way different communities, frequently with very different traditions, have contended with similar social problems, and the comparative features of Social Anthropology, Psychology, Law and Sociology provide some of the most instructive and useful aspects of these disciplines. Currently, there are a number of problems which engage the attention of social scientists and administrators. They are not new problems; but they appear to have acquired a degree of urgency over the last few decades.

In the years of economic depression in the early 1930s, a common observation was that the world was 'overproduced', a view probably fostered by reports of coffee being used as fuel for railways, of grain being destroyed or not planted to ensure price levels, and of millions of unemployed without the purchasing power to acquire the goods available. The view of the Rev. Thomas Malthus (1766–1834)— *Essay on Population*, 1798—that eventually the world's population

would outstrip the capacity for the production of food, a view which
influenced Charles Darwin in the formulation of his theory of natural
selection, seemed irrelevant or a distantly remote possibility; but by
the 1950s the proposition was commonly seen to be of more than
academic interest. During these years, and the trend has continued, it
also became more widely realised that the habitat of many animal
species was being either spoiled or destroyed so that many animal
species were confronted with interruptions to behaviour patterns
essential to their life cycle and the possibility of their extinction. The
present century has also seen the emergence of large, socialistic
republics in which the private ownership of property and many of
the exclusive rights associated with territory and property are restricted, thereby challenging an important incentive in the motivation
of man and many animal species. Convictions for crimes involving
aggression and drug-taking have increased in many countries but
may reflect, in part, improved methods of detection and recording
and while Utopian methods of retreat from, or attempts to solve,
social imperfections are not new, 'drop-out' cults from the obliga-
tions of social life are now much publicised by modern mass media.
Societies have today, and have always had, problems. In an era of
rapid communication, there is probably a greater awareness of the
problems of societies and of their complex nature and a genuine interest in how different societies contend with them. Factors deriving
from different climatic, historical, cultural, demographic and techno-
logical backgrounds of human societies render the various solutions
interesting enough; but it is possible to introduce another dimension
of interest by considering how different species with different behavioural resources contend with some of the basic problems of survival;
because of the field studies of many ethologists and some genuinely
comparative psychologists, it is now possible to do this.

If a species is to survive and be effective, there are certain features
which must be ensured. The species must be capable of reproducing
young which are in turn capable of rearing young who will become
effective parents. Many facets are involved. An adequate food supply
must be available which involves that, within the habitat, the numbers of the several occupying species are adjusted so that the environment is not exploited beyond its regenerative capacity. More immediately, an effective method of dispersion is necessary and the dispersion, as will become apparent, has social implications in addition to
the animals' nutrition. Many species also require a period of attachment between mother and young not only for survival of the young
but also in order that important aspects of the mothering process,
probably of an emotional kind, may be known to them. Female
goats, sheep and Rhesus Monkeys which have been reared artificially

have in many cases proved to be ineffective mothers. Members of the species must also be able to find their way about. Without adequate direction-finding abilities, many species could not complete their life cycle or, as in the case of bees, birds and bats, for example, their daily round of activities.

Many species also require a stable social order for which forms of communication are necessary. This very often means an order based upon a dominance hierarchy which is helpful not only in preventing aggression from rising above wasteful and destructive limits, but also ensures that the young have an opportunity to become acquainted with the facts of social life, with rights, concessions and appeasement and the methods of communication appropriate to each. There is evidence, too, that the successful manipulation of sexual functions and the appropriate methods of approach and courting are acquired not only by life within the animal community but also during the sexual phases of juvenile play which the protection of an organised society makes possible. It is also probable that some aspects of the care of the young are learned within the social group. The advent of cubs in a pride of Lions or infants among a troop of Chimpanzees, Baboons or female Langurs is often associated with improved relationships within the group. Even old, male Baboons are tolerant of the young and on occasion may carry them or allow them to climb on their bodies or to ride pick-a-back. Again, as noted by Hediger (1962), Goma, the baby Gorilla born in the Basel Zoo, was nursed upside down by his mother, who had been reared by human attendants. The many Rhesus Monkeys reared on cloth-covered, wire-frame mothers by Harlow and Harlow (1961) were once regarded as a potential breeding stock, but were useless as such because of their ineptitude in sexual approaches. So, too, the Burmese Red Jungle Fowl cocks, reared in visual isolation by Kruijt (1962, 1964), had no other method of approach to the hens than violent aggression with far less than optimum reproductive success. Through deprivation of social experience, the cocks were deprived of the normal postures and communications of courtship.

Further, the social order of many species would not be stable without some viable distribution of territory—in effect, a satisfactory degree of dispersion. As already emphasised, a degree of dispersion is essential for continuing and adequate nutrition; but it is also necessary in many species for successful reproduction. Inadequate spacing inhibits many forms of social communication, including those associated with mating. In many species of birds and mammals, mating is associated with the establishment of a territory by the male, without which very little mating would occur. Encroachment upon the habitat of animals should thus be assessed not only in terms of potential nutritive loss but also in terms of disturbance to behaviour patterns essential to survival.

The emphasis in the present work is upon field studies of animal behaviour, a bias which has no evaluative reflection upon complementary laboratory studies, which can very often pose questions in controlled and specific form. In defence of field studies, it may be claimed that whereas many experimental studies in the laboratory can better explore the limits of performance in any specific activity, field studies probably give a more reliable account of what the species has achieved within the complexities of the ecology of its environment. Again, it should be remembered, as Agar (1943) pointed out, that there is not only the possibility of selection of organisms by the environment but also a selection of environments by organisms.

The possibility of the operation of a form of conscious awareness of the environment and of the organism's own needs and related conscious, purposeful behaviour by the organism has confronted every student of animal behaviour; but the indication of the stage of evolutionary development at which it may be claimed to operate is a profoundly difficult task. Despite all the dangers of anthropomorphism, of attributing human attributes to animals in excess of their powers, the operation of conscious awareness and related, purposive behaviour cannot be shelved. To include them renders the explanation of behaviour the most difficult field of scientific explanation but some position in regard to these problems is demanded. They will be considered against the background of the aspects of behaviour which were indicated above as essential for the survival of species; but first, some difficulties in the process of explanation itself must be considered.

2

DIFFICULTIES OF EXPLANATION

THE PRINCIPLE OF ASSOCIATION

Despite all the difficulties of perception, of knowing the properties of things, uninfluenced by cultural background, habit and personal bias, there can be little doubt that change is typical of external reality and of life itself. It is difficult to conceive of effective behaviour or of the survival of any organism which has no power of registering and reacting to change. There is empirical evidence of such reactions at rudimentary levels of evolutionary development. Mast and Pusch (1924) were able to show that, when part of the *Amoeba* comes into strong light, the number of attempts to continue in the same direction decreases with successive trials. They refer to this change as *learning* (their italics); but it should be noted that in a later paper, Mast (1932), accumulation of gelatin under the stimulation of light, with consequent increase in the elastic strength of tissues in the stimulated area, led to contraction in this area and bulging in others. Viewed externally, the change was adaptive; but a thorough materialist could argue that everything could be explained by reference to the concepts of Physics and Chemistry and that there was no need to import any properties of consciousness or purpose into the process, as a Vitalist would tend to do. Nevertheless, there are a number of other observations which are a challenge to explanation. Mast and Hahnert (1935) demonstrated that *Amoeba proteus* was selective in its 'choice' of food. Jennings (1906) describes how an *Amoeba* appears to 'chase' a cyst of *Euglena*, the pseudopods advancing repeatedly in an apparent movement of enclosure.

At a more complex level of evolutionary development, Yerkes (1912) was able to show that the worm *Allolobophora foetida* could be trained to turn to one side of a 'T' maze in which one arm led to a moist, dark enclosure (conditions favoured by worms) and the other to an abrasive surface where contact was made with irritants and/or electric shocks. It was found that with marked individual differences, the habit of turning to the favourable side could be established in 20 to 100 trials. Other observations, which are applicable to human learning, were that spaced trials were more effective than massed trials and that performance improved with increasing trials. Here is evidence of some rudimentary and slow-working power of association of events. Whether or not it is in any way linked with some crude form of awareness, is impossible to establish; but at a far distant level of evolutionary development, namely that of human behaviour, the property of association is fundamental to human thought. Indeed without association, thought would be impossible and so too perception, knowledge, and explanation. That which is perceived is related to something else which is known. In what may be regarded as a limiting case, we may be unable in darkness or in fog to know anything more than that there is 'something over there'. We are unable to give any details of its size and nature. We merely know that it is something other than ourselves and 'out there'; but a relationship exists. We are aware of it and we may be said to know something. Clearly, the meaning derived from each new experience will be influenced by the existing knowledge of the person which in turn will reflect the previous experience and the pattern of society and culture in which they have lived.

In a comparable way, explaining involves a process of relating either of concomitant variations which are observed or relating the novel experience to that which is familiar in past experience. There appears to be some abatement of curiosity or provocation, once some relationship has been indicated; but obviously the criteria of satisfactory explanation will vary according to the sophistication of those concerned. Remarks like 'something to do with the new power scheme' may initially reduce curiosity over the appearance in the locality of strange, new machinery; but for some people, further processes of relating would be necessary to ensure full abatement of tension or curiosity. Broadly, there are two classes of variables involved in the process of explanation. First, explanation is an activity carried out by processes of human thought which reflect many developmental and cultural factors. Secondly, the processes which are to be explained differ markedly both qualitatively and in complexity, so that it is defensible to claim that some processes are easier to explain than others. Some of these problems are now reviewed.

DEVELOPMENTS IN HUMAN THINKING

It would seem to be natural that the child's first attempts at explanation should be of the animistic type where the chief causal factor is an animated, i.e. a human or quasi-human, agent, such as the goblins or fairies which cause unfavourable or favourable events to happen. The rain-gods or thunder gods of mythology are in the same category. For in early infancy, the first and most effective causal agent which gets things done in response to the child's clamant demands is a human agent and when events take place in external nature, it is natural that the same form of explanation should be used. Logically, it has certain distinctive features. The animated agent is in imagination held to be responsible for causing the event and producing the observed result. The implication is that the effect is ensured by the agent or that there is some necessary connection between cause and effect. But the development of the early conceptions of causal association takes place within a wider context of activity.

The poet Wordsworth in the prose introduction to *Ode on the Intimations of Immortality, from Recollections of Early Childhood* mentions that in early childhood he often had difficulty in thinking of himself as detached from or other than the environment with which he was familiar:

but there may be no harm in adverting here to particular feelings or experiences of my own mind on which the structure of the poem largely rests. . . . But it was not so much from [feelings] of animal vivacity that my difficulty came as from the sense of the indomitableness of the spirit within me. I used to brood over the stories of Enoch and Elijah and almost to persuade myself that, whatever might become of others, I should be translated, in something of the same way, to heaven. With a feeling congenial to this I was often unable to think of external things as having external existence, and I communed with all I saw as something not apart from, but inherent in, my own immaterial nature. Many times while going to school have I grasped at a wall or tree to recall myself from this abyss of idealism to reality. . . . To that dream-like vividness and splendour which invests objects of sight in childhood, every one, I believe, if he would look back, could bear testimony. (*The Poetical Works of Wm Wordsworth*, Ed. E. de Selancourt, Clarendon Press, Oxford, 1947.)

This sense of mergence of oneself with the environment has been studied by professional psychologists who have related this attitude or mental set both to perception and explanation. If one, by contrast, is not feeling merged with the surroundings but very much detached from an environment which is regarded as 'out there' and 'other than' oneself, one might expect that perceptions of the external world and reproductions of it would tend to be more objective, dis-crete and articulate. Werner (1940) and his co-workers have, on the other hand, illustrated the comparatively diffuse character of child-ren's drawings of actual objects and situations. A pyramid was re-produced with several apices because the child felt it to be a 'pointy' thing and the sharp 'pointiness' was the vague, syncretic impression which was fundamental for the child. The author has observed a five-year-old child who, when asked to draw the teacher on the black-board, seized a piece of chalk and struck the point of the chalk several times against the board to the accompaniment of the exclama-tion ugh! ugh! ugh!. It was in no sense an articulate reproduction but in a syncretic way, an effective one. Piaget (1926a, pp. 61 *et seq.*) has suggested that syncretism is also associated with other attitudes of mind. One of these he terms 'nominal realism' and his description is: 'In learning the names of things, the child at this stage believes it is doing much more. It thinks it is reaching to the essence of the thing and discovering a real explanation.' Any teacher will confirm that this is a tendency not confined to very young children. Naming of a phenomenon, with no further knowledge than the mere name, is often accepted even by unsophisticated adults in lieu of an explana-tion. The mere process of naming, perhaps by suggesting some form of relationship by linguistic association, can be associated with some reduction in provocation; but a further consequence of syncretism in the child is seen by Piaget as follows (1926b, p. 185):

he has a tendency, both in verbal intelligence and in perceptive intelligence (and the tendency lasts longer in the former than the latter type of mental activity), to look for a justification at any cost of what is either simply a fortuitous occurrence or a mere 'datum'. Now in verbal intelligence this tendency to justify at any cost is connected with the fact that the child thinks in personal, vague and unanalysed schemes (syncretism). . . . These schemes connect with one another all the more easily owing to their vague and therefore more plastic character. In this way, the syncretism of verbal thought implies a tendency to connect everything with everything else, and to justify everything . . . [and further (1926a, p. 212)] for behind the most fantastic events that he believes in, the child will always discover motives which are sufficient to justify them; just as the world of primitive races is peopled with a wealth of arbitrary intentions, but is devoid of chance.

Both Werner and Piaget have indicated, as did Wordsworth, on the basis of his own impressions, that syncretism and animism tend to decline with increasing mental age; but the several forms of animis-tic explanation recorded by Piaget indicate the potential ramifications of these attitudes of mind. In many cultures, the animistic type of explanation of many natural events tends to linger. When the anthropologist Malinowsky (1923) studied the people in the Trobriand Islands, off the north-eastern coast of New Guinea, in 1914–18, the prevalent explanation of birth and death followed an animistic pattern typical of many myths. Sexual copulation between husband and wife made it possible for the spirit of a dead chieftain to enter the womb of the woman. The spirit came from an offshore island where it had been resting, either on driftwood or on the wings of birds and was the real progenitor of the child. As in the accounts of the births of Venus, Romulus and Remus, Moses, Lohengrin, and Scyld Scefing in *Beowulf*, birth was associated with arrival by way of water.

In many cultures, however, economic reality soon exposes the limitations of animistic explanations. If avalanches or floods destroy villages or crops, appropriate siting and the building of retaining walls and dykes are found to be more successful than placatory ceremonies to quasi-human agents. Again, as the child gains experience of the operation of natural causes, such as the force exerted by wind and water, the understanding of non-animistic explanations and the need to master the environment results in an increasing tolerance and use of causal explanations in which the causes are natural events that can be observed and readily associated with observable effects. Nevertheless, the early history of man's acquaintance with the causal relationship is often evident. Many communities may display great ingenuity and engineering skill in building and irrigation and yet for events over which they have no control, such as violent electric storms and rainfall, they may adopt deistic, i.e. basically animistic, explanations in which the chief causal instigator of events is a quasi-human agent and therefore presumed to be amenable to prayer and supplication.

The development of the concept of causation in the individual is naturally much influenced by the cultural background of the community in which he lives; but the wide range of development in the conception of the causal relationship may also be illustrated by some reference to the history of science.

There is ample evidence that many of the earliest attempts at explanation were animistic in character. In ancient Egypt the apparent course of the sun across the sky was explained by attributing the phenomenon to the passage of a glowing chariot driven by a deity. Even the great Newton, whose laws of motion formed the basis of a major step towards an era of objective science, could claim (*Principia*, 'Gen. Scholium', Book III) that 'the most beautiful system of the sun, planets and comets could only proceed from the council and dominion of an intelligent and powerful being'. Newton more than once expressed his dislike of hypotheses; but despite this and his enormous services to objective science and facilitation of accurate prediction of a wide range of natural events from tidal movements to eclipses and other astronomical events, he needed an animistic assumption to get his system going and readily accepted an animistic and deistic conception of creation, for as he wrote (*Optics*, 4th ed., p. 400):

> God in the Beginning formed matter in solid, massy, hard, impenetrable moveable Particles, of such Sizes and Figures, and with such other Properties, and in such proportion to Space, as most conduced to the end for which he formed them . . . and therefore, that Nature may be lasting, the Changes of corporeal Things are placed only in the various Separations and new Associations and Motions of these permanent Particles.

Newton, as may be observed of many scientists today, could be objective for the greater part of his scientific thinking and revert to an earlier stage of explanation in others.

The association of animism with the earliest stages of explanation, both in the life of the individual person and in the history of science, may be at least in part responsible for other features in attempts at causal explanation. One of these is the marked tendency to conceive of or demarcate a process into 'cause' and 'effect', a dichotomy which might appear to be somewhat arbitrary and may derive from a certain narrowing or focussing of attention because of personal needs or interest. The 'effect' is often of direct concern to the individual making the explanation, and the 'cause', in Collingwood's (1937) phrase, tends to be 'the handle whereby we manipulate the event'. Kelson (1941, 1943) has suggested that this fundamental dichotomy may in fact derive from an early stage of animistic thinking, in particular that associated with the 'lex talionis' or law of retribution as encountered in tribal society. Action against another tribe tends to be associated with a reaction or reprisal. To act against another tribe or individual leads to feelings of guilt and the expectation of a reaction of comparable severity. This concept of action and reaction is the basis of the old legal conception associated with the early Hammurabian code, of an eye for an eye or a tooth for a tooth. Inevitably, in a culture where animistic explanations are common, social norms or what is normal in social events tend to become associated with events in physical nature and thus the dichotomy of 'cause' and 'effect' arises.

Also deriving from the animistic beginnings are the features of causal necessity and a suggestion of a more general, all-pervading legal or moral order associated with the notion of universal causality which may in turn be associated with the concept of a 'prime mover'. Newton appears to have held this point of view and indeed much of the epoch-making scientific work of the sixteenth and seventeenth centuries could have been done and probably was done against a background of such assumptions.

The most notable direct attack on this position came from the Scottish philosopher David Hume (1711–1776) whose writings were claimed by Kant to have awakened him from his dogmatic slumbers. It would appear natural, having regard to the animistic origins of the concept of causality and man's need to control some aspects of his environment, to regard the 'cause' as *producing* the 'effect' and indeed to conceive of the effect as following of necessity from the cause. But Hume's writings oblige us to reflect and to appreciate the possibility that *production* of the effect by the cause and any suggestion that the effect must necessarily follow the cause, or that any features of the effect are prescribed or entailed by the nature of the cause, may be primitive, vestigial remnants of an earlier era of thought. For Hume, despite certain minor inconsistencies in presentation, there could be no element of production or entailment.

We have no other notion of cause and effect [Part III, Sect. VI] but that of certain objects, which have always been conjoined together, and which in all past instances have been found inseparable. We cannot penetrate into the reasons for the conjunction. We only observe the thing itself, and always find that from constant conjunction, the objects acquire an union in the imagination . . [and again (Part III, Sect. XIV)] . . there is no internal impression which has any relation to the present business, but that propensity, which custom produces, to pass from an object to the idea of its usual antecedent.

The causal relationship is thus reduced to a habit of mind. We become so accustomed to rubbing cold hands 'to make them warm' that we imagine that the rubbing 'produces' the warmth. We know too, from other experiences, that friction is associated with heat; but Hume obliges us to consider that the relationship between friction and heat is no more than a statement of probability. In all our past experiences, friction has been followed by or associated with heat. Therefore, on a future occasion, the same association is highly probable. The probability may approach but cannot equal certainty. There is the possibility that on some future occasion, friction may not be associated with heat. The strict empiricist, who can only be guided by what has been observed, must adhere to the facts and to import 'production' and 'necessity' into the causal relationship is to go beyond the facts. Later logicians might argue that something is entailed between cause and effect; but the Humean view of the causal relationship is definite and there are further implications. If one adopts the Humean view, one cannot speak of anything 'producing' a result or indeed of ensuring a particular result or event, or of such everyday occurrences as the new medicine making grandfather better or the fertiliser improving the growth of the lawn. One would only be justified in observing that the application of the fertiliser was associated with increased growth and the same linguistic strictures would be relevant if particular phases of the chemical processes involved were to be considered.

Actually, effective scientific work can be done whether or not causal relationships are thought to involve necessity. If one is interested in the relationship of the expansion of a rod of a particular metal with rise in temperature, that relationship can be accurately represented by recording measurements of length and temperature and arranging them on a graph. So, too, the relationship of the volume of a given mass of gas and the pressure exerted upon it at a constant temperature can be graphed. The relationship of concomitant variation can also be expressed mathematically, indicating how one variable is a function of the other. Predictions can be made and the relationship can be manipulated by a person from the graphical or mathematical summary. Debate on the existence of necessity in the relationship is of genuine psychological and logical interest, but not essential for the purposes of accurate description, prediction and control. Provided that three fundamental features, spatial and temporal continuity and some form of concomitant variation or functional dependence between supposed cause and effect can be demonstrated, useful scientific work can be done. It is quite conceivable that large numbers, possibly the majority of scientists, work quite successfully without concern over problems of necessity in the causal relationship and that many adhere to the natural and pre-Humean viewpoint that the 'cause' produces the 'effect'.

However, a possible further stage in the progression of attitudes to the causal relationship may be associated with the advent of the study of intra-atomic processes. Hitherto, reasonably accurate prediction of an event had been associated with a certain confidence that if the cause or causes of an event were known, accurate prediction was possible. The study of intra-atomic events, that is, events involving several small particles moving at very high speeds, emphasised the need for statistical approximations of the probable state of an atom at particular times. In this context, the statistical association and prediction is more realistic and useful than any considerations of the 'cause' or 'a cause' or 'the causes' of the state of the atom at any point of time, even though individual scientists may experience a certain lack of conviction and yearning for some indications of determinism.

DIFFERENT TYPES OF PROCESSES

While there is ample evidence that attitudes to explanation have varied in different historical epochs and may vary in a Western culture from person to person, another important group of variables are associated with the way processes differ in themselves. From a psychological point of view, some types of process are easier to explain than others. Consider first a simple case:

(1) We place a book on a smooth desk and gently push it along the surface, a distance of, say, one foot. The apparent 'cause' of the movement of the book, which is the apparent 'effect', is movement of the hand. There is thus a form of qualitative similarity or qualitative continuity between apparent cause and effect. Again there is apparent temporal and spatial continuity between cause and effect. There is no challenge or sense of hiatus to human understanding which requires the insertion of some construct between apparent cause and effect or indeed of some intervening condition. Movement is observed to be the 'cause' of movement and if appropriate allowances are made for frictional resistances, no energy is mysteriously lost. To unsophisticated human observation, which would probably not involve the issue of 'necessity' between supposed cause and effect, the whole event, which is familiar in everyday experience, presents the minimum of provocation. Further, in addition to the features of spatial and temporal continuity of apparent 'cause' and 'effect', the functional dependence of 'cause' and 'effect' can be easily demonstrated. The speed of movement of the book is dependent upon the forces applied to it and a graph or mathematical expression of this function would be readily possible. The simple 'causal' pattern of explanation has an apparent relevance to such simple events.

(2) Consider another example; we take medicine in the hope of relieving a condition of digestive discomfort. A few days later, we become aware that the discomfort has gone. Colloquially we say that the medicine has 'done its job' or that we are 'cured' and regard the taking of the medicine as the cause of the cure. In contrast with example (1), there are some important differences. There is a sense of qualitative dissimilarity between apparent cause and effect and also of temporal and spatial discontinuity between cause and effect. Naively, we suspect that 'something has gone on' or that some other processes have intervened between apparent cause and effect. Primitively, we find gaps of this kind difficult to tolerate and feel a desire to fill the gap by making clear, causal sequences in the processes

which we imagine must have taken place between the somewhat arbitrary end-observations, namely the taking of the medicine and the awareness of the cure. We hypothesise and insert hypothetical, intervening variables between our observations to render more plausible our account of what may have happened during the process, and, if well formulated, such hypotheses may be tested.

(3) A similar sense of hiatus is apparent in the next example. In the breeding season, the male three-spined stickleback will court not only the female, which prior to the emission of ova has a distended ventral surface, but also crude plastic models with a distended under-surface. It will also attack other males which at this time have a distinct reddish colouration of the under-surface of the body. Several comparable examples could be cited from nature. The chick of the Herring Gull, for example, pecks at a red spot on the lower mandible of the mother's bill. She regurgitates a morsel of food onto the ground, picks it up in the bill and offers it to the chick. The responses take place initially, without evidence of learning. The stimulus of the distended shape of the female or the red colour appears to impinge upon some innate organisation which is already arranged to mediate the response characteristic of the species. Wm McDougall (1936) postulated an intervening construct, an *instinct* which he defined (p. 25) as 'an innate psychophysical disposition which determines its possessor to perceive and pay attention to objects of a certain class, to experience an emotional excitement of a particular quality upon perceiving such an object, and to act in regard to it in a particular manner'. Some animals were claimed to possess several such instincts or innate patterns of reaction to specific stimuli and each one was thought to be mediated by a specific 'neurological correlate' or pattern of neurones. For many years after its first publication in 1908 in *An Introduction to Social Psychology*, the inserted construct of the instinct had little more than an inferred, hypothetical status; but in 1943, W. R. Hess and Brügger were able to show that by passing small electrical currents to particular parts of the hypothalamus, a small central area of the brains of intact cats, by means of fine inserted electrodes, they were able to evoke specific patterns of behaviour such as fighting, eating and sleeping. The cats not only ate and went to sleep but searched for food and for a place in which to go to sleep. Later work by von Holst and von St Paul (1960) with inserted electrodes in the brains of domestic cocks has lent support to the hypothesis that specific innate patterns of response to perceived features in the environment may be mediated by specific organisations of neurones.

Later workers than McDougall, namely Lorenz and Tinbergen, have used the inserted construct of an *Innate Releasing Mechanism*

(IRM) as an aid in explaining the consistent patterns of response of animals to specific stimuli. The many consistencies in the patterns of behaviour which are distinctive of a particular species and which ethologists have discovered appear to call for some inserted construct between the observation of stimulus and response, and it is possible that the loci of the relevant neural organisations will in future be more adequately mapped.

(4) A casual glance at any good newspaper will reveal that in many parts of the world large numbers of people care or feel strongly on a number of issues and can easily be provoked to violence. But one need not go to the politically troubled areas of the world to find evidence of patterns or organisation of emotional reaction. One could be reasonably confident of evoking a hostile reaction by going among a section of partisan supporters at a football match, disparaging the efforts of their side and applauding the opposing side. A mere symbol or a chance remark has been known to provoke a riot. Again, from our knowledge of a person we know well, we can frequently predict with accuracy his attitude on many issues, that certain of his duties will be meticulously discharged or that violence would be highly improbable.

Common sense would suggest that an enormous amount of 'shaping' or organisation of emotional activity occurs between infancy and adult life. Indeed the nature of this process is the topic of a major part of psychological literature and the nature of the organisation is the source of many hypotheses involving inserted constructs and intervening variables. McDougall (1936) used the concept of a *sentiment* which involved the focussing of one of the innate emotional dispositions such as fear or anger or a fused emotion such as scorn (in his view, a fusion of anger and the innate emotion of disgust) about an object or idea. These individual sentiments later became integrated under a *self-regarding sentiment* which, in the light of the several emotional attachments, represented the individual's conception of himself in society as he would have himself.

Allport (1947) used the concept of *trait* to represent the enduring result of the impingement of experience upon the innate drives to particular forms of action. He defined the trait (p. 295) as 'a generalised and focalised neuropsychic system (peculiar to the individual), with the capacity to render many stimuli functionally equivalent, and to initiate and guide consistent (equivalent) forms of adaptive and expressive behaviour'. Traits were specific to the individual and were 'adverbial' rather than 'prepositional', that is, they determined the individual's way of doing things rather than, as in the case of an attitude, a particular orientation towards an object or viewpoint. Traits, as Allport conceived them (p. 319), 'on the physiological side,

are undoubtedly neural dispositions of a complex order'. They had
to be inferred from statistical evidence of the person's behaviour
since, at the time Allport wrote, convincing neurophysiological
evidence of such organisations was lacking, although today it appears
less unlikely.

Clearly, the sense of hiatus, or the impression that many processes and features of organisation must have intervened between the
perception of a situation and the behaviour subsequently observed,
is considerable. Compared with example (1), the process is much
more complicated and the issue arises as to whether or not the causal
pattern is applicable to explanations of human behaviour. In view of
the relevance of past, personal experience, historical, geographical
and socio-economic factors to the behaviour of the individual and
groups of human beings and the individual differences deriving from
genetics, it is clearly impossible to specify all the factors which are
spatially and temporally continuous with and functionally related to
a particular act of behaviour.

A theoretical case can be made for the view that the experience of
a given situation is not identical for any two persons, since the meaning and significance of every facet of the situation derives in part
from the previous experience of the individual and yet despite the
multiplicity of factors, no therapy, rehabilitation or even education
of higher animals or human beings would be possible if behaviour
were entirely chaotic and devoid of all presumption of causal consistency.

SPECIAL DIFFICULTIES OF EXPLAINING BEHAVIOUR

The behaviour of living organisms, however, presents particular
difficulties for explanation because of the large number of specialised
scientific disciplines which are relevant. In order for successive acts of
behaviour to take place, sequences of neurophysiological, physiological, chemical and, at some levels, conscious processes must occur.
Several procedures are possible. One is to distinguish, say, only
physiological changes and explain such changes by reference to the
resources of physiological science. Another is to try to estimate
which physiological and chemical processes are most relevant and
include an account or estimate of their contribution in the attempted
explanation. The physiological and chemical effects of prolonged
frustration or tension could, for example, be included in an explanation of disturbed thinking or displacement of emotional behaviour.
Another approach would be the 'reductionist' procedure in which
conscious mental states were 'translated' or represented by the
neurophysiological processes which mediate them and these in turn

are further 'translated' into the presumed, more basic physical and chemical processes underlying the neurophysiology. One problem is the possibility of loss, that all features may not be 'translatable' at each stage and also that new features may emerge.

There is also a certain and possibly somewhat arbitrary presumption that the form of a question constrains the answer or the explanation, that a problem posed in chemical terms, for example, should be explained in chemical terms or that a sequence of conscious states should be explained in terms of experience. A certain prescription by the terms of the question is difficult to avoid and technical difficulties impose limitations; but the chief determinant of such excursions from one subject to another will always be human interest or provocation and this may derive not only from purely scientific interest but also from practical expedience. A bad case of truancy or sibling rivalry will involve many acts of behaviour with physiological, neurophysiological and chemical aspects; but it may prove more useful for purposes of therapy to seek the explanation in the past experience and behavioural history of the persons concerned. So, too, the ineptitude of the Harlows' Rhesus Monkeys (1961) and Kruijt's Burmese Red Jungle Fowl (1964) in sexual approaches might more usefully be associated with their restricted early experience, unless this important loss of social learning can be adequately represented in physiological or chemical terms.

WAYS OF AVOIDING THE DIFFICULTIES

Some methods of by-passing these difficulties have been used in the past. They are:

(1) *The giving of reasons as an explanation of human conduct*
This is a frequently used method of explanation and one much used in legal proceedings where it is associated with systems of jurisprudence. If we are asked why we came to town such answers as 'I came to see my doctor' or 'to change my library books' or 'to meet some friends who were passing through town' would be commonly accepted as explanations of the behaviour of a reasonable man. Explanations based on human motives, being basically animistic and teleological in character, present less of a challenge in a field where human motives are presumably involved. There is the general difficulty, suggested by psychoanalytic literature, that on occasion reasons sincerely given may be mistaken; but the indication of a common type of motive consistent with one's usual mode of life has, if only broadly, sufficient in common with other forms of explanation to render it acceptable. One could assume that the need or motive

corresponds very roughly to the 'cause' and the subsequent behaviour to the 'effect', that the cause and effect are, in a sense, spatially and temporally continuous with one another and that there is some form of functional dependence between the intensity of the motivation and the energy expended in the behaviour. But the correspondence with the causal type of explanation is only broadly approximate. The need to change one's library books can be met in many different ways so that it would be difficult to claim consistency or precision in the sense of invariable association between a particular need or desire and the subsequent behaviour.

Wm Dilthey (1894) and others of the 'verstehende' school argued that the only defensible way of explaining human behaviour is by an empathic process based upon a complete knowledge of the person's history—such as would be conveyed by the statement, 'I have known him and his background for a very long time. I had an intimate knowledge of the circumstances leading up to his behaviour on that day and I know exactly how he must have felt before he assaulted the man.' Many novelists and playwrights have used this method with convincing results. Meredith's *The Egoist* is a good example and so too, in the social sphere, is Chekov's *The Cherry Orchard* which portrays more effectively than the work of many a social scientist the impression of a society devoid of *élan* and grinding to a halt. The English philosopher Collingwood (1936) held a view broadly similar to that of Dilthey. To know the operation of minds, the best method is to study history. 'For history, the object to be discovered is not the mere event, but the thought that expressed it.' The cause of the event for the historian 'means the thought in the mind of the person by whose agency the event came about' and the only way in which the historian can discern these thoughts is 'by rethinking them in his own mind'.

The Italian philosopher Benedetto Croce (1941) has also drawn attention to the potentialities of empathy associated with a knowledge of historical events when he pointed out that, for brief periods, an historical novelist may succeed in conveying to the reader more convincingly than the professional historian just how men and women thought and felt at the time. But while such methods may on occasion succeed in providing a true reconstruction of past experiences and 'verstehende' methods may be of value to the clinician in assessment and therapy, the evidence gathered by Meehl (1954) indicated that in prediction of future success in occupations, statistical prediction, based upon objective test measurements, are more reliable than clinical methods where a personal knowledge of the patient is involved. In view of the enormous, potential range of human personalities, the degree of presumption in judging the experience of other persons by one's own is evident; yet without the assumption of some degree of communality, literatures would be meaningless and translation from one literature to another impossible.

(2) *Another possible solution is behaviourism*
The observer may elect to proceed solely on the basis of observations of behaviour which is regarded as an event 'out there'. There will be no mention of aspirations, intentions, feelings or any other form of experience. Any consistent tendencies will be evident only as a result of statistical counting based on events 'out there'. Any conjunctions or linkages of different phases of behaviour will again be based on statistical counts only, and usually in the form of correlations. The whole procedure will be basically empirical. In dealing with animal behaviour, the possibility of anthropomorphism, the attribution of human qualities to animals, is much reduced and in the case of human beings, the attribution of false motives is less probable. While conceivable in theory, there are some genuine difficulties, in particular those associated with the language in which the behaviour is to be described. The use of such words as 'endeavour', 'elude', 'avoid', 'pursue', 'try', 'plan', 'contrive', etc. is questionable because, as a result of their normal usage, the operation of conscious purposeful activity is suggested and a strictly behaviourist programme would entail that, if such words are used, they must be defined in a strictly behaviouristic way, in terms of observations 'out there' which several observers can confirm. Anger, for example, would have to be defined in terms of specific gestures, colouration of the skin, intensities of sound and possibly heart-rate and chemical composition of the blood, assuming that recording of these indicia is compatible with the observation of continuous behaviour in a real-life situation. One obvious result would be a great change in the language used in describing behaviour and possibly, a change of emphasis deriving from the fact that some aspects of behaviour have more obvious, external manifestations than others. However, while these general difficulties apply to all forms of behaviourism, it is relevant to distinguish some different forms of behaviourism which have arisen or are possible. These include:

Type (a): A completely empirical science based upon the whole organism 'out there', i.e. behaviourism at the molar level with no presumptions about the nature or purpose of life or that behaviour, is purposive or otherwise. This would of course be subject to the methodological and linguistic strictures mentioned above.
Type (b): An extension of the above empirical method to include physiological and neurological states where particular states were

demonstrably associated with observed behaviour. This method would also encounter the difficulties mentioned above.

It is in this general classification that the basic approach of *physicalism* arises. As presented by Carnap (1959, p. 165), 'every sentence of psychology may be formulated in physical language'; and again, 'all sentences of psychology describe physical occurrences, namely the physical behaviour of humans and other animals'. All psychological concepts, it is held, may be defined in a way which derives from physical concepts and psychological laws may be translated into the same language. The procedure is governed by the verification principle, namely that (p. 174) 'a sentence says no more than is testable about it'. If in describing or attempting to explain a man's behaviour, we use such terms as depression, excitement, or anger, implying that they are inner states and cannot clearly present the objective observations of physical phenomena which would verify the existence of such states, then we are, in Carnap's view, indulging in metaphysics. The statement by a reliable person that he was or felt depressed, excited or angry would be of no value unless observable phenomena could be indicated which would verify them and these could be observations of inner physiological states or outwardly observable behaviour. At first acquaintance, this procedure would appear to have certain merits of objectivity and avoid some well-known difficulties. Even if Jones and Brown are reliable, and familiar to the observer, there is no indication from their reports that a state of anger in one is qualitatively or quantitatively identical with that in the other, or indeed how, with any exactitude, they may be compared. Ordinary conversation and literary usage proceed on the implicit assumption that such words as anger, resentment, depression and sadness do refer to states in different persons having something in common; but this is an assumption probably reinforced by repeated overall, broad impressions of consistency in the statements of different persons about themselves and their behaviour in the external world.

There are, however, further difficulties. In judging and stating that another person is, say, excited, an observer may be guided by inflexions of the voice, gestures and other outwardly observable indicia. These, in principle, could be supported by records of pulse rate and chemical tests although these would introduce an artificial element and probably a departure from normal social behaviour. But these indicia are 'out there' and, as such, verifiable by more than one observer. When an outwardly undisturbed man quietly concedes that he is sickened by the trend of events at a meeting or that he currently suffers from digestive discomfort or headaches, the statement is not based upon observations of events 'out there' and the method of verification cannot be the same as in procedures applied to others. As Malcom (1964) has indicated, the content and method of verification of statements made in sentences in the first person (i.e. of oneself) and in the third person (of others) is different and they cannot with consistency be combined.

In the same general class of behaviouristic approaches is that of B. F. Skinner (1953) for whom behaviour is a function of some condition which can be described in physical terms. These conditions may be internal or, preferably, external and behaviour is viewed as a variable dependent upon these conditions which have the status of independent variables. These independent variables must be carefully defined. To claim that an animal drinks because it is thirsty is merely a circular statement of the kind that an animal migrates because it has a tendency to migrate and adds nothing to knowledge or explanation. If, however, some dimensions could be indicated, such as hours of food deprivation in the case of hunger, this condition can be linked with the behaviour which is dependent upon it. Assuming that a condition of deficiency exists, an action which mediates a reduction of the deficiency, such as a movement which secures food, may be said to be reinforced and will tend to be repeated. In this way and basically comparable ways, a considerable patterning of animal and human behaviour can be achieved which extends far beyond simple, nutritive conditions to cultural and social conditions where positive and negative reinforcement, in the form of social approval and disapproval, can modify behaviour. By reinforcing positively the operations of advantage to the organism an extensive range of behaviour can be acquired; but the resources of the Skinnerian forms of behaviourism come under strain when applied to some forms of apparently purposive behaviour. 'Instead of saying that a man behaves because of the consequences which *are* to follow his behaviour', wrote Skinner (1953, p. 87), 'we simply say that he behaves because of the consequences which *have* followed similar behaviour in the past'; and again (p. 89), 'In general, looking for something consists of emitting responses which in the past have produced "something" as a consequence'.

With human beings, general knowledge, deriving from the use of language, results in few situations being completely novel. A modern city dweller cast up on a desert island would, like Robinson Crusoe, try something in the general direction likely to produce results and follow up lines of activity which were positively reinforced. A rat or pigeon in the confined space of a Skinner box requires an appreciable time before its behaviour is shaped or constricted to press the lever, which produces the reinforcement. Once some form of linkage is established, however, between lever and reinforcement, search for the lever as a means to an end is understandable in any animal with some power of retention and capacity for independent and variable behaviour. More difficult to include within the same conceptual scheme are the feats of navigation which result in Manx Shearwaters, for example, flying alone from Boston or Venice to Skokholm off the south-west coast of England, a very long journey not hitherto undertaken by the individual member of the species, or the successful navigation by bees in sunlight on directions acquired the previous day in shadow when clues to the sun's position were mediated by the plane of polarised light.

The wide application of the principle of reinforcement is nevertheless emphasised, but even in the constricted conditions of a Skinner box some initial exploratory activity is necessary before the lever is encountered, and some power of retention on the part of the animal must be assumed.

Type (c): A very interesting form of behaviourism was propounded by the late Professor E. C. Tolman (1932) who wrote (p. 416) 'our task, *as psychologists,* is the collecting and ordering of the *molar behaviour* facts *per se*. And this task can, in large part, be performed in relative ignorance of physiology and neurology.' But there were other important assumptions underlying Tolman's programme. On the basis of his observation, he could write (p. 12):

. . . there seems to be no other way out. Behaviour, as behaviour, that is as molar, *is* purposive and *is* cognitive. These purposes and cognitions are of its immediate, descriptive warp and woof. It is, no doubt, strictly and completely dependent upon an underlying manifold of physics and chemistry, but initially and as a matter of first identification, behaviour as behaviour reeks of purpose and of cognition.

Another feature distinctive of Tolman's approach is (p. 12) that: 'such purposes and cognitions are just as evident, as we shall see later, if this behaviour be that of a rat as if it be that of a human being'.

Tolman's problem is thus that he recognises that behaviour is purposive and involves cognition, or awareness and knowing; but if he is to use the behaviouristic method consistently, he cannot speak of what the animal 'thinks' or 'feels'. He takes the view that 'raw feels' 'do not get across' to the scientific observer; but what do get across to the observer of behaviour, in addition to the situation and the apparent stimuli, are certain 'behaviour determinants' which function as intervening variables between the initiating physiological state and the observed behaviour. They are frequently not directly observed, but must be inferred by observers. A rat whose heredity, training and state of food-deprivation is known can be observed at the conjunction of several pathways to move from one to the other, perhaps entering each for a short distance, before finally proceeding along one of them, which may lead to food. On another occasion, the rat may be observed to use certain ledges as footholds to climb around an obstruction in the maze or to drag out bundles of straw in which pellets of food are embedded. Several independent observers could confirm that the rat made a discrimination and had manipulated the straw prior to seizing the food. In Tolman's view, the situations contained a 'discriminandum' for the rat, a 'manipulandum' and 'a means-end relationship'. An environment can thus be regarded as consisting of a number of such variables which are to be inferred from observable behaviour and from his point of view (p. 414) 'there is nothing private or "mentalistic" about them. They are pragmatically conceived, objective variables, the concepts of which can be altered and changed as proves most useful'.

The student of Tolman's interesting system becomes aware that the general underlying conception is very similar to that of Wm McDougall (1936). What McDougall would call Instinct, i.e. 'An inherited or innate psychophysical disposition which determines its possessor to perceive and pay attention to objects in a certain class, to experience an emotional excitement of a particular quality upon perceiving such an object, and to act in regard to it in a particular manner', Tolman provided for by such inferred variables as 'Means-End-Readiness', 'manipulanda expectations', 'discriminanda expectations', 'Sign-Gestalt-Expectations', 'demanded type of goal' and others. Very often the rat established the discriminanda within a situation after a period of running to and fro and 'vicarious trial and error behaviour'.

The hesitation which a reader may feel in assessing the system derives from the realisation that much behaviour must be observed in the same situation and a very high level of consistency observed before the inferred variables could be regarded as having any reliability. Again, while the claim, that much thought tends eventually to be manifested in action, must be seriously considered, and much human thinking can be presented as analogous to 'mental trial and error' and an appraisal of strategies, the task of attaining statistical reliability of the different manifestations will be seen to be very lengthy and complicated and likely to lag behind the mental development of the individual and perhaps, too, the cultural changes of the community, be it animal or human, in which it lives.

Type (d) The Hullian behaviouristic system: Any appraisal of the methodological resources of psychology should mention the attempts at system-building of the late Professor Clark L. Hull, outlined in three works, *Principles of Behaviour* (1943), *Essentials of Behaviour*

(1951) and *A Behaviour System* (1952). The guiding principle in these three works derives from Hull's conception of explanation which was (*Principles*, p. 2) that 'an observed event is said to be explained when the proposition expressing it has been logically derived from a set of definitions and postulates coupled with certain observed conditions antecedent to the event'. In his view, the essential feature of 'scientific theoretical explanation is that it reaches independently through a process of reasoning the same outcome with respect to (secondary) principles as is attained through the process of empirical generalisation' (*Principles*, p. 5). Admittedly, there are many satisfying features associated with attaining the same result in different ways. If, from certain simple definitions of a Euclidean triangle, it can be deduced that the three angles of a triangle add up to 180° and an extensive programme of measuring of the angles of triangles gives a corresponding generalisation, i.e. if deductive reasoning from first 'principles' and empirical investigation give the same result, the standing of this result is enhanced.

Again, one could conduct experiments and show that the product of the volume and the pressure of a given mass of gas at a constant temperature is constant and it would be possible to deduce the same result from the assumptions of the kinetic theory of gases. So, Hull embarked upon an extensive programme which he described (*Principles*, p. 2) as 'the beginning of a systematic objective theory of the behaviour of higher organisms', a theory being defined by him as 'a systematic deductive derivation of secondary principles of observable phenomena from a relatively small number of primary principles or postulates much as the secondary principles or theorems of geometry are all ultimately derived as a logical hierarchy from a few original definitions and primary principles called axioms'.

Hull began with simple, neurological findings dealing with the passage of energy in neurones, neural interaction and the establishment of habits and related these to concepts of 'drive', 'inhibition', 'stimulus generalisation', 'reinforcement' and 'behavioural oscillation'. Methods of quantification and inter-relation of units followed the methods of Physics. The entire system was arranged on the basis of sixteen postulates in the *Principles* (1943), extended to eighteen in the *Essentials of Behaviour* (1951) and refined to seventeen in *A Behaviour System* (1952). Only relatively simple learning behaviour, based upon the assumption of 'drive' and the concept of reinforcement, was covered; but ultimately, Hull hoped to cover the whole range of behaviour. As he wrote (*Principles*, pp. 25–6), 'an ideally adequate theory even of so-called purposive behaviour ought, therefore, to begin with colorless movement and mere receptor impulses as such and from these build up step by step both adaptive behaviour and maladaptive behaviour. The present approach does not deny the molar reality of purposive acts (as opposed to movement), of intelligence, of insight, of goals, of intents, of striving, or of value; on the contrary, we insist on the genuineness of these forms of behaviour. We hope ultimately to show the logical right to the use of such concepts by deducing them as secondary principles from the more elementary objective primary principles.'

It was a bold plan and an enormous amount of experimental research was devoted to discovering, refining and checking findings which could be built into or derived from postulates; but the important issue is whether or not an extensive, deductive system could be developed in Psychology. It is conceivable that prolonged and careful study of movements of animals in a stable and customary environment will indicate patterns of movement which could be designated as purposive or insightful. Ethologists have at least indicated such possibilities; but if such observations are to provide the basis of a deductive system, the behavioural events conjoined must provide an invariable sequence of the standing of a law and it is questionable, even with the most consistent instinctive patterns, if observations of animal behaviour would provide the degree of consistency required. The problem would appear to be more difficult at the level of human behaviour where conscious attitudes such as intentions and the many forms of striving would have to be conjoined in the invariable manner of a law with particular movements. In view of the potential range of human values and tastes, the difficulties are enormous. Even the simple desire to go to town could be associated with wide variations between persons and with the same person from time to time.

Type (e) The use of models: Lord Kelvin (1884, pp. 131–2) remarked that he only completely understood a process if he could make a model of it. Conceivably, if a model provides only an approximate analogy, it has some explanatory force in the sense that it may reduce human curiosity or provocation; but even if a cybernetic model can achieve the same final solution to a problem as an individual animal or human subject, it does not necessarily follow that the two processes of solution are identical. Models in general offer an hypothesis as to possible ways in which the behavioural event may have occurred and may be explained. Like analogies, they have to correspond very closely to the event to have explanatory value and this is exceedingly difficult if the variety of events in the distant and immediate past and anticipated future, which may contribute to behaviour, are to be represented adequately.

B

PREVIEW

From the foregoing brief analysis, it will be apparent that behaviour of all animals, including man, presents considerable difficulties in explanation. Apart from the affinities suggested by the use of the same language in dealing with the behaviour of widely different species, it has frequently been implied in psychological literature and particularly since the writings of Charles Darwin (*On the Origin of Species*, 1859, *The Descent of Man*, 1871, *The Expression of Emotions in Animals and Man*, 1872) that animal and human behaviour have much in common. The whole of McDougall's system, the most comprehensive to date, for explaining animal and human behaviour was based upon the implicit assumption of the continuity of evolution in behaviour. Apart from some gaps, the evidence for continuity in evolution in bodily functions is impressive and perhaps some indication of possible continuity and affinities in behaviour and the property of purpose may be gained by referring to the solutions which members of some species with different behavioural resources have attained with basic problems encountered in their life cycles.

As indicated in the introduction, the study will tend to range beyond the laboratory to studies in the field and because of limitations of space will concentrate on some aspects of behaviour which are important for survival. These are attachment of mother and young, the ability of species to find their way about and some features of the social order of different species.

There is, however, one further recurring difficulty in making any comparisons between species to which reference has already been made. In describing and evaluating human behaviour, it is commonly assumed that the behaviour is purposive and derives from a conscious awareness of features of the environment. There is thus no difficulty in using such words as 'strive', 'endeavour', 'pursue', 'restrain' or any other words, which in centuries of usage and in translation from one language to another have been employed to describe the behaviour of human beings when assumed to be behaving purposively; but to use such language to describe the behaviour of animals is questionable for there is a risk of being indirectly anthropomorphic, [of attributing to an animal by implication forms of behaviour which may differ from what it actually may have]. But the basic problem, admitting that there may be differences of degree, is to decide at what point in evolutionary development purposive behaviour is possible.

Tentatively, and as a guide in the ensuing treatment, the features listed below are suggested as indicative of purposive behaviour.

Independence: The organism is not at all times a passive thing moved by environmental forces, but has some capacity to resist or behave independently of these forces.

Variability: There may be variation in the nature and direction of successive acts of behaviour until a result which is consistent with the previously recorded behaviour of the species is observed. The same behavioural result may also be attained by the same animal in different ways on different occasions and at times in different ways by different members of the same species.

Retention: Behaviour can persist independently of the situations which may have instigated it, or evidence that behaviour is influenced by objects or situations not immediately present.

The case for purposive behaviour is stronger if the behavioural event can be shown to involve all three criteria.

While it must be recognised that a cybernetic device could be designed with all of these features, it could be claimed that the device would be incorporating features of purpose or purposiveness, supplied by the designer.

3

PROBLEMS OF ATTACHMENT

The attachment of human mother and young is sanctified by a long social, religious and literary tradition although the bases of attachment are by nature very strong. Unlike the young of many non-mammalian species, the human infant obtains its early nutrition directly from the mother by suckling, a process which is in the great majority of cases pleasant and rewarding to both mother and young and thus reinforcing to all aspects of attachment. It is further reinforced by the protection and feelings of solicitude afforded to the young. The almost complete helplessness of the neonate is in itself a compelling feature, without the great weight of tradition; but cases of child neglect are recorded and parents have not infrequently been obliged by law to provide the minimum conditions of parental care. In several societies over the centuries infanticide or abandonment has been practised, particularly in difficult times, and currently many parents in western civilisations are prepared to relinquish their offspring to foster parents. The effectiveness of the child-parent bond in other species thus acquires greater interest.

ATTACHMENT IN BIRDS

The achievement of attachment of mother and young in many species without the aid of tradition must nevertheless be effective if survival is to be enhanced and perhaps the most striking method is that indicated by the processes of imprinting which may be observed quite clearly in several avian species where the young are precocial and nidifugous, i.e. capable of movement at or very soon after birth and therefore likely to leave the nest soon after hatching. The most familiar examples are provided by domestic chicks, ducklings, and goslings which soon after hatching are capable of following the mother about and will continue to do so for several weeks. The young do not obtain food directly from the mother, but by following her the probability of obtaining appropriate foods and of protection against predators is enhanced. The term imprinting derives from the German verb 'prägen' = to coin or stamp—and the related noun 'Die Prägung' = stamping—and appears to have been used first by Oskar Heinroth (1911) with reference to his observations of the large family of waterfowl (Anatidae) of which certain members, notably newly hatched geese and ducks, were seen to respond to the first large moving object which they encountered, very much in the same way as they would normally react to their parents. Human attendants, for example, were persistently followed and in some cases became the object of sexual advances by the birds in later life.

While the use of the term imprinting is rightly associated with the name of Heinroth and the development of his insights by Konrad Lorenz, observations of the phenomenon have been recorded over a long period. Pliny the Roman encyclopaedist (*Natural History*, Book X, p. 51) having referred to the geese which gave warning of an attack on the Capitol mentions a particular goose which became attached to the philosopher Lacydes, 'never leaving him, not in public, in the baths, nor by day or night'. Watt (1951) mentions that Reginald, a medieval monk of Durham, in 1167 noted instances of Eider ducklings following human beings; but it was Douglas Spalding, an Englishman, who died in 1877 in his late thirties, who provided the first scientific insight into the phenomenon. In a paper before the British Association in 1872, published in the following year in *Macmillan's Magazine*, Spalding reported a number of cogent observations. As soon as they were able to walk, domestic chicks would follow objects moving across his table. Those fitted with small blindfolds would follow his hand when these were removed at the age of three days. There was a brief period of hesitation during which the chicks emitted the familiar high-pitched 'fear' chirps, but very soon they were following his hand. Those unhooded on the fourth day, however, fled from him with every sign of fear. Spalding concluded that this reversal of the chicks' behaviour must be due to changes which had occurred within the chick, and in so doing drew attention to the operation of a 'critical period', during which responses of following and attachment most readily take place, and to the emergence of fear with which the termination of the critical period has been associated by more recent workers.

In later studies of the phenomenon, Lorenz (1935, 1937) came to the view that imprinting was characterised by certain distinctive features, namely that:

(1) The process is confined to a definite, brief period in the early life of the individual.

(2) Once established, the process, unlike the more familiar, associative learning, is irreversible.

(3) The imprint may be generalised in the sense that the animal tends to react to the broad or general characteristics of the species or indeed the object upon which it may be imprinted.

(4) The process of imprinting may be established long before it is evident in behaviour. Lorenz (1937) has mentioned a male Muscovy (*Cairina moschata*) hatched with four siblings by a pair of grey geese which the drake followed for seven weeks. In all social activities, the bird associated with his own siblings; but later when his mating responses began, they were addressed to the species of the foster parents which he had not seen for ten months.

As will be shown later, subsequent research would indicate that (2) might be relaxed somewhat; but it can be pointed out that, while most studied in birds, phenomena broadly analogous to imprinting have been reported of non-avian species.

Noble and Curtis (1935, 1939) reported that male, cichlid fish, which were reared in isolation for the early part of their life, addressed mating reactions to either sex. Baerends (1950), in following up these findings, recorded observations on the species *Cichlasoma meeki*, on *Hemichromis bimaculatus*, which spawn on the bottom, and on *Tilapia natalensis*, which incubate the young in the mouth. The young of *Hemichromis* did not show the fright response during the first two days of free swimming. By the use of models which varied in size, shape, texture, colour and type of movement, some specific reactions were found. The fish moved away from stationary or rapidly moving models. Shape did not appear to be crucial; but size was, in that the young fish followed at a distance which consistently subtended the same angle with the model. Lorenz has reported analogous findings during the imprinting of young goslings when, by crouching, he varied the height of his head above the ground. *Hemichromis* showed a marked preference for red; but in *Cichlasoma* the preference for red was not evident till after the age of 15 days. *Hemichromis* also appear in some way to retain the impression of places and objects previously encountered. When an area of the sand on the bottom which contained spawn was removed, the female returned to the same area and aerated for about half an hour before the response waned; but it was resumed when imitation eggs were inserted. Eggs of another species, however, disappeared and so, too, the young of other fish (presumably devoured), whereas parents retrieved their own young.

The remarkable ability of bees to learn rapidly the type of pollen and the direction and the distance of its location by following other bees in a dance (see von Frisch, 1967) is also broadly analogous to the discrimination associated with following which is characteristic of imprinting in birds. The acquisition of song patterns by some species of birds also has some features analogous to those found in examples of imprinting.

With the Chaffinch (*Fringilla coelebs*) Thorpe (1963) reports that birds reared in isolation produce very simple songs, apparently corresponding to what are the inborn and basic components of the song. When such birds are brought together, the eventual pattern of the song is very abnormal. It appears that in the early weeks of life, the young birds in normal circumstances acquire from parents and other birds singing nearby the essential features of the Chaffinch song, such as the division into phrases, the range of pitch and the conclusion with a flourish. Thorpe indicates that there is good evidence that this learning is possible both in the wild and in the laboratory without immediate practice by the young bird; but in the first breeding season, the bird sings in competition with other birds for territories and in so doing learns many of the finer details and ornamentations of the song.

More recent research has suggested that the basic conceptions of what is involved in imprinting might well be re-examined. Basically, the animal which is to be imprinted must do two things. It must make a movement of approach to the imprinting stimulus or object and, if this is a moving object, it must continue to approach, i.e. follow, and it must be capable of discriminating the particular imprinting stimulus or type of stimulus from alternatives. As James (1959) and Smith (1960, 1962) have shown, relatively simple, intermittent stimuli such as a flickering (on-off) patch of light will attract domestic chicks. Smith also demonstrated that movement in the frontal-parallel plane (as across the surface of a wall, viewed straight ahead) was a significantly more attractive stimulus than movement in the sagittal plane, i.e. directly ahead, as one would fire an arrow. Again, objects or stimuli moving at eye-level for the bird are more attractive than those above eye-level, 20° being approximately the critical angle of regard. Smith and Bird (1963b) with domestic chicks have shown that simultaneous presentation of auditory and visual stimuli are more attractive than visual stimuli alone and (1964c) that response tends to be better in those chicks which encounter the auditory stimulus first. The hereditary component in response to visual stimuli (Smith and Templeton, 1966) does not appear to be very considerable. The correlation between the approach scores of parents and offspring was not significant; but it was positive and the best estimate of the probable contribution from heredity was in the neighbourhood of 18 % of the variance, using the same stimulus conditions and methods of housing in successive generations. This would suggest that factors operating in the egg, at hatching or between hatching and testing, are important.

However, attraction is only one aspect of the process. An animal which more than once approaches or follows an object is in a position to learn some of the characteristics of the object, in other words to acquire discrimination, and the crucial question is whether the discrimination is acquired over a period of time in a manner comparable with ordinary associative learning and increasing familiarity, or quite suddenly, which would put the process of acquirement in a special category. It may be that species differ at this point. Both Smith (1962) with domestic ducklings (Khaki Campbells) and Klopfer and Hailman (1964) with Peking ducklings (*Anas platyrhynchos*)—the domestic form of the Mallard—have shown that with sufficiently attractive stimuli the birds will transfer their preference from a plain to a more colourful stimulus. Both of these studies also found no convincing evidence for the 'hypothesis of effort', suggested by Hess (1959), that the strength of imprinting was directly proportional to the effort expended in following. Further evidence that perceptual variables are very important is provided by the study of Smith and Nott (1970) which showed quite clearly that, with a combination of auditory and visual stimuli which previous experiments had shown to be very effective, it was possible to obtain a virtually maximum score in approach and discrimination with a domestic chick which first encountered the stimulus at the age of ten days post hatch, i.e. long after the so-called sensitive period was thought to have terminated. With a smaller number of Mallard ducklings it was also possible to prolong the critical period of responsiveness as far as approach responses were concerned; but dramatic differences occurred in discrimination. The Mallards appeared incapable of discriminating consistently between the familiar and unfamiliar stimulus, a result which directs attention to the distinction made by Lorenz (1935, 1937) between the 'mosaic' and 'regulative' type of imprinting which, subject to further research, could perhaps be taken to represent two extremes of a range.

In the 'mosaic' type of imprinting the young bird would appear to have some innate 'schema' or image of the parent and will attach itself to that specific stimulus pattern and no other. Lorenz cites as examples the newly hatched Curlew (*Numenius arquata*) or a Godwit

(*Limosa limosa*) which flee at the sight of a human being. Very differ-ent is the 'regulative' type response of the Greylag gosling (*Anser anser*) which in Heinroth's (1911) observation must be shrouded when emerging from the incubator, if it is not to become attached to the experimenter. It may be that evolutionary selection and domesti-cation have contributed to these differences and conditions of hatching and the strength of the imprinting stimulus may be addi-tional variables. Smith and Harding (unpublished) have shown that rewards of food and water will increase the approach scores of domestic chicks which are already high, so that the approach without external rewards or reinforcement may be said to be extensible by conventional types of reinforcement.

Again, Mallard ducks and domestic chicks do form mild attach-ments to stationary objects left in their pens over long periods and tend to follow them when they are moved. Klinghammer and Hess (1964) have reported imprinting in altricial birds (those helpless at birth and in consequence, nidicolous, i.e. spending a long period in the nest). Squabs of Blond Ring Doves (*Streptopelia risoria*) remain in the nest for about fourteen days. They estimate the optimum time for imprinting to be seven to nine days. Immelmann (1967), by interchanging eggs between clutches of the grass finches, Zebra Finch (*Taeniopygia guttata castanotis*), Bengalese Finch (*Lochura striata f. domesticata*) and the African Silver Bill (*Euodice cantans*), allowing the young to be raised by foster parents and then isolating the young birds when they had attained independence, found definite evidence of both social and sexual imprinting. The males preferred females of the foster parents or models thereof. They were later found to pair with females of their own species if no choice existed; but the preference for the foster species still remained, in some cases for the several years of testing. Immelmann thinks nevertheless that imprinting on the bird's own species is stronger and less readily re-versed than on foster parents and, in this sense, favours an innate bias or the influence of some 'schema'.

Clearly, the whole basis of imprinting could be re-examined and the role of the several relevant variables more clearly defined. No doubt, precisely controlled laboratory studies will be necessary; but some further indication of the complexity of problems of attachment and their place in the life cycle of the species will appear from the studies discussed below.

Hinde, Thorpe and Vince (1956) working with young Moorhens (*Gallinula chloropus*) and Coots (*Fulica atra*) found that a wide variety of moving objects, including a walking man and a large canvas 'hide', would elicit avid following and most birds revealed a power to generalise, i.e. to transfer their following response from one

moving object to another. The essential feature appeared to be movement. Fabricius (1951) working with Tufted Ducks (*Aythya fuligula*),
Eider Ducks (*Somateria molissima*) and Shoveler Ducks (*Spatula
clypeata*) found that the shape of his models did not seem to be very
important; but motion was essential, especially if parts of the model
moved with reference to one another, as happens when the parent
bird moves. In this context, the contrary action of the legs of a man
while walking might provide a very attractive stimulus complex.
Weidmann (1958) noted, too, with Mallard ducklings that they
would not approach a motionless human being; but did so if the
experimenter moved or swayed from side to side. He noted, as did
Fabricius, that the introduction of short, low-pitched sounds were
very helpful. Weidmann also noted that ducklings isolated for 40
hours would not follow the model, but those isolated for as long as
50 hours would move to join other ducklings, or if they had not seen
other ducklings, they would join newly hatched Moorhens and
Pheasants. This might suggest that while the young bird during the
sensitive period will approach a range of moving objects, some repeated or continuing experience of the object is necessary before discrimination develops. Fabricius (1951) on the basis of his work with
young Tufted Ducks, Eiders, Shovelers and Mallards came to a
similar view, that these species did not have an innate image or
'schema' of the parent bird, in effect that discrimination and attachment was in part at least due to a process of learning.

Many workers including Lorenz (1935), Nice (1953), Collias and
Collias (1956), Hinde, Thorpe and Vince (1956), Weidmann (1958),
Pitz and Ross (1961) have noted that auditory stimulation has a
facilitating effect on initial, approach responses of young birds.
Collias and Joos (1953), in a careful analysis with the sound spectro-
graph, established that for the domestic chick the most attractive
sounds are of low pitch, i.e. below a frequency of 800 cycles per
second with the best responses occurring between 49 and 392 c/s. The
most effective presentation is an oft-repeated and well segmented
pattern of such frequencies, with each note of short duration. While
detailed findings for several bird species are not available, it is frequently noted that attractive calls tend to be low-pitched and of short
duration, while alarm calls tend to be high-pitched and more protracted.

Klopfer (1959a,b) reported that the surface nesting species,
Mallards and Redheads (*Aythya americana*), if reared in auditory
isolation, will approach a variety of low, rhythmic sounds apparently
without discrimination. Wood Ducks (*Aix sponsa*), a hole-nesting
species, however, soon developed a tendency to approach a specific
pattern of sound to the exclusion of others, i.e. to imprint auditorily.

Muscovy Ducks (*Cairina moschata*) gave a poor response to sounds. Shelducks (*Tadorna tadorna*) were found to develop a preference for particular sounds if the sounds were associated with a visual model. These differences could reflect an element of evolutionary selection. In the restriction and semi-darkness of the nesting-hole, there are probably greater advantages if the young imprint on the soft 'kuk kuk' calls of the mother which in the Wood Duck begin as soon as the eggs are 'pipped' and increase in intensity, as do the ducklings' calls, as the time of leaving the nest approaches. The mother eventually drops to the surface of the water and begins calling. One after another, the ducklings climb to the rim of the nesting hole which is often in a hollow of a tree and tumble down, a distance which has been recorded as high as 60 ft. When no more sounds issue from the nest, the duck swims away, ahead of the brood.

There is good evidence that stimulation at the foetal stage may influence later behaviour. Gottlieb (1965) has indicated that in natural circumstances the head of the duck and chicken embryo enters the air space at the large end of the egg some days before hatching and, soon afterwards, the embryo begins to peep intermittently at low intensity. During this period, the incubating duck is known to call at low intensity. Gottlieb further studied the problem by inserting fine electrodes into the lower part of the bill of the embryos of White Rock chicks and Peking Ducks on the day before hatching and recorded any vocalisations on a very sensitive microphone. Another electrode under the left wing provided evidence of heart action. Maternal calls of the species were played on tape at about 70 decibels for periods of 30 seconds. The embryos always responded by beak clapping and vocalisation between bursts of recorded sound. In further experiments, Gottlieb (1966) showed that White Rock chicks and Peking ducklings which had been visually isolated after hatching, but exposed to their own and the noises of other newly hatched birds, followed models emitting adult calls of their own species in preference to other models emitting juvenile calls. In another group, exposure to recordings of juvenile calls was associated with improved response to adult calls of their own species, but not to the adult, exodus call of the Wood Duck.

A later study (Gottlieb, 1967) distinguishes between physiological activation of the embryo (increase in heart rate) and behavioural activation (increase in oral activity). On the day before hatching, the Peking duck embryo shows behavioural activation only to the maternal call of its own species. On the other hand, physiological activation occurred in response to the maternal calls of several species. Behavioural activation to the maternal call was on further study found to occur as early as three days before hatching. It was observed that on the fourth day before hatching, many of the embryos began to vocalise and, to test the relevance of this peeping, eggs at this stage were placed in individual, auditory isolation and tested three days later, i.e. on the day before they would normally hatch. The embryos isolated in this way did not respond behaviourally to the maternal call of the species, though they did show physiological activation. The exposure to the 'peeping' of embryos in nearby eggs would appear to influence the embryos' ability to respond selectively to the maternal call of its own species.

Of some interest in this context are the studies of Vince (1964, 1966a, b) who has recorded two rhythms from eggs of different species of quail, a hatching rhythm of sounds of relatively large amplitude and 'clicking' which sets in some 20 to 12 hours before hatching. The number of clicks per minute appears to differ for the eggs of different species of quail, e.g. the Bob-White Quail (*Colinus virginianus*), 82–160, and the Japanese Quail (*Coturnix coturnix japonica*), 61–196. By placing eggs at different stages of incubation into clutches of other incubating eggs, Vince found that while the timing of pipping of the egg was not affected, the interval between pipping (first rupture of the egg by the bill) and hatching could be reduced, presumably because of stimulation from rhythms in nearby eggs. A later study indicated that stimulated eggs tended to hatch first and that eggs in contact hatched within the space of six hours. Isolated eggs hatched over longer periods—46 hours with the Bob-Whites and 63 hours with the eggs of the Japanese Quail. For the quail, survival is clearly enhanced if the period of immobility of hen and brood is reduced and mutual stimulation between the several eggs of the batch would appear to make an important contribution.

In very different circumstances, Carr and Hirth (1961) have noted a further example of the importance of early social stimulation. In studies made in Costa Rica, it was observed that the young of the Green Turtle (*Chelonia mydas*) have a far better chance of emerging from the nest in the sand and reaching the sea if the numbers are large. Single young are rarely successful; but with larger numbers of eggs, between 50 and 100, the success approached 100% in the absence of predators.

The relative contributions of innate preferences, early stimulation and facilitation and later perceptual learning would obviously merit further study; but clearly there are advantages for survival if, by any means, the young of the species can discriminate and become attached to a larger and more experienced animal which in turn is capable of discriminating particular young and assists in their nutrition and protection. For a mobile species or one nesting in relative isolation, such an arrangement would on balance probably be more effective and economic of energy than all adults participating indiscriminately in the care of any young of the group and there probably are long-range advantages for the young as will become apparent in studies yet to be mentioned. The degree of discrimination can in some species be quite impressive.

Sladen (1958) reported of the Adélie Penguins (*Pygoscelis adeliae*) that at about the age of four to five weeks, the young are left on the land in large groups or crèches while both parents are at sea obtaining food. It was observed that the returning parent walked steadily through the crèche, while all the young watched intently. Recognition of parent and offspring appeared to be mutual and was sometimes marked by loud displays by both. During the Antarctic summer of 1948–9, coloured rings were fitted to the flippers of chicks of marked parents before the crèche stage. Of 71 observations, only twice were parents found to be feeding chicks other than their own and these were actually in addition to their own. Close study would probably determine the perceptual cues be they visual, auditory, tactile or olfactory or combinations thereof which mediate this order of discrimination; but it does seem to be effective and the young survive in what must be one of the most rigorous environments in the world.

ATTACHMENT ON OTHER SPECIES

Some of the most effective attachments of parent and young are to be found among the mammals and particularly between mother and young where the influence of food reinforcement is probably considerable; but the bases of discrimination are of interest.

The newly born of sheep, goats, moose, horses, camels and domestic cattle will approach the nearest large object in the vicinity and there is some evidence that the response takes place more readily if the object embodies some movement, as the mother usually does. Pruitt (1960) and Lent (1966) have in fact noted the head bobbing of the Caribou cow (*Rangifer tarandus*) which appears to be helpful in inducing the calf to follow. Smith (1965), Smith et al. (1966) were able to study closely the lambing of the domestic breeds of sheep, Mule (Hexham Leicester ram × Swaledale ewe), Half-breed (Border Leicester ram × Cheviot ewe) and Clun Forest. A common first indication that a birth was due was the withdrawal of the ewe for varying distances from the flock and the sniffing of the grass upon which the amniotic fluid had fallen after the rupture of the amnion. Experience showed that it was very difficult to keep the ewe away from this area. If driven away, she frequently managed to return to it and here, after labour, the lamb was born. Ewes which had lambed in previous seasons were usually quicker to respond to the lamb. Those having their first lamb were often slow to react; but the sound of the wet, struggling lamb and in particular its first, feeble bleating usually attracted their attention and the licking and nuzzling of the neonate began, accompanied by short, repeated phrases of a low-pitched bleat. The lamb thus received tactile, auditory and visual stimulation from a moving object which had definite olfactory characteristics. The ewe was receiving gustatory and olfactory information which corresponded with the amniotic fluid previously encountered on the grass and it is defensible to speculate that the process of learning of the olfactory characteristics of the lamb begins when the ewe begins to sniff the grass. Conceivably, the probability of the ewe finding the lamb could be enhanced in this way.

Normally, the lamb has struggled to its feet within twenty minutes of parturition and with its early faltering steps an innate reaction is observed, namely, to place the head under any object or projecting surface at about the height of the lamb's shoulder and under which it can thrust its head (Fig. 1). This proved to be the fold of trousers overhanging the observer's knee boots, the lower wooden rail of a fence, an old overcoat or corn sack folded over a trestle, the upper side of a box, the ledge of several models and *any* part of the under-surface of the ewe. The thrusting movements of the head frequently took the form of butting, accompanied by the shaking of the tail which is typical of lambs when suckling. The finding of the teat and an effective orientation to it was the result of much trial and error, extending in some cases over an hour, with thrusting and butting movements under many inappropriate regions of the ewe and even sucking on wisps of wool.

The process of attachment was further studied by contriving that the lamb was born onto a disposable polythene sheet and the ewe moved immediately to a clean pen in a barn sufficiently distant from that in which the lambs were housed to ensure that bleating could not be heard. By returning different lambs at progressively longer intervals after birth, Smith et al. (1966) reached the tentative conclusion that for the breeds studied, the ewe so deprived of the early opportunity for nuzzling would accept any lamb, and not necessarily her own, up to at least eight hours after parturition. The time interval was not extended further to avoid possible damage to the lambs. It is possible that the ewe may have been influenced by confinement in a pen where her own odours would tend to predominate, and by deprivation of the natural tendency to nuzzle; but within the eight-hour period, any lamb which the ewe was able to nuzzle for 20 to 30 minutes was allowed to suckle by the ewe consistently maintaining a nursing orientation. No more than two lambs were presented to

Fig. 1
The innate tendency is for the lamb to place its head under any object about the height of its own shoulder.

any one ewe and one at least was not her own. In all, 37 lambs were returned to 21 ewes with only one case of immediate refusal and this lamb was accepted 35 minutes later. A comparable result, though only in the short term, was reported in a later study by Klopfer and Klopfer (1968) who showed with Toggenburg goats that an alien kid presented to the doe immediately after parturition was as likely to be accepted as the doe's own kid (13 out of 15 of the aliens were accepted compared with 11 out of 15 own kids). Moreover, if the doe had already accepted an alien, she would also accept one of her own kids. The doe and the ewe appear to be disposed to accept a variety of young immediately after parturition, although if presented with a *choice* might prefer their own. As shown later, the doe's acquisition of an olfactory imprint appears to be very rapid.

Smith et al. (1966) also observed the role of visual and auditory cues. Sheep have good vision, to judge by their power to anticipate and avoid capture when the experimenter approached from a distance, downwind; but the ewe does not appear to make great use of her visual powers in recognising her young. Auditory cues at a distance appeared to be more important. By recording the bleating of individual lambs, Smith (1965) was able to show that ewes which had been accepting their lambs for four or five days would reveal agitated searching and bleating at the absence of one or both lambs and, despite the many competing noises of the flock in a closed space, could find the tape recorder hidden around several corners and partitions in the barn. Grabowski (1941) reported a similar finding of auditory discrimination by a young hand-fed lamb which came to distinguish the voice of its human 'nurse' from those of others; but this could be a clearer case of rewarded or reinforced learning.

At close quarters, olfactory cues appear to be most important. Young lambs are avid 'opportunists' and will suckle greedily from any ewe who will maintain a nursing orientation, even for a few seconds. Suckling lambs are, however, almost invariably sniffed by the ewe and 'scroungers' either butted away or the nursing orientation is not maintained. The failure to encounter a nursing orientation would in natural circumstances be fatal for the neonate and there is some evidence that the critical period during which the mother forms an impression of her own young may in some species be very short, as shown, for example, by Klopfer et al. (1964) with Toggenburg goats. One group of 14 does ('immediate separation group') were deprived of their kids as soon as they were born and had no opportunity to lick them. Another group were allowed 5 minutes of nuzzling and licking of the new-born kid which, if interrupted by the birth of another, actually ranged up to 10 minutes, but usually not more than 6 minutes. Kids were later presented to the does at intervals of

1, 2 or 3 hours after birth. Of the 'immediate separation' group, 12 out of 14 rejected their kids. Of the 'prior contact' group, only 2 out of 14 rejected—statistically, a very significant result. A period as short as five minutes was found to be crucial and blindfolding had no effect. The importance of olfactory cues in these attachments would appear to have been confirmed in later experiments by Klopfer and Gamble (1966) when the corresponding figures for the 'immediate separation' group were 13 out of 14 rejecting and 15 out of 17 accepting for those allowed 5 minutes of contact. In a further experiment, 6 does had a spray of cocaine inserted inside their nostrils before the birth. This did not lead to rejection of their own young in acceptance trials, but did seriously interfere with discrimination of their own from other kids who would normally be rejected. When in another experiment, the spray was applied after the does had licked the kids and before the acceptance trials, 4 out of 7 does rejected their own kids, possibly because of a failure to obtain a matching impression. It was also noted that whatever the reaction in the tests, it persisted when the effects of the cocaine had worn off, a result which might be due to perseveration or inertia on the part of the ewe or hesitancy on the part of lambs which had once been rejected.

Dramatic though the attachment may appear to be in the above experiments, it can frequently be overcome by slow habituation. It is an old shepherd's trick to secure the head of a ewe to prevent butting and allow the lamb to suckle until eventually it is accepted. Hersher et al. (1963) by this method cross-fostered kids on ewes and lambs on does between 2 to 12 hours after parturition. In an average of 10 days, all of the adoptions were established with the cross-species, foster mothers reacting more vigorously to separation.

The foregoing studies emphasise the role of the mother in the process of attachment and, in sheep, it appears that she plays an important and probably dominant part; but a study by Alexander and Williams (1966) of fine wool, Merino sheep suggests the operation of some innate tendency in the lamb. By fitting covers over the teats, they established that the proportion of time spent in teat-seeking declines from about 30 % over the first 3 hours of life to less than 5 % at the age of 12 hours. This early peak is probably not entirely due to hunger, since much teat-seeking was observed after giving milk to the lambs. Again, hunger might be expected to increase with the passing hours whereas teat-seeking declined. Further study will help clarify the contribution of each party which probably varies between species; but Klopfer's finding that a period as short as five minutes can be sufficient to enable the doe to form an effective discrimination and attachment suggests a possible interpretation of an observation by Blauvelt (1955) that, as soon as the newborn kid becomes mobile, the mother tends to maintain a 'safe' area around it by butting intruders away—in effect reducing interference and ensuring favourable conditions for the attachment of the kid to the doe which in a mobile, grazing species has obvious implications for survival.

Lent (1966) has reported a comparable development between Caribou cow and calf. Agonistic acts are directed towards strange calves and yearlings that approach too closely and some effort in the form of head bobbing, grunting and nuzzling by the cow is apparent in strengthening the bond between the mother and the calf which exhibits the familiar generalising tendency to follow any moving animal. In moments of danger to the herd, this generalising tendency could have definite survival value for the neonate. A similar observation was recorded by Altmann (1958) of the Moose (*Alces alces*). The calf does not become fully mobile until about the fourth day when its marked tendency to follow or 'heel' the cow, initiated by the cow advancing and waiting, and also other moving animals or a man on horseback, is apparent. As the days pass, the radius of the calf's ranging distance from the mother increases, but it is also evident that the cow is maintaining a 'safe' area around the calf by driving other animals from the area and the calf back into the area, when it has strayed. The relationship between the Moose cow and her calf may reflect the relatively solitary life of the Moose. The calf is still orientated towards the cow during the rutting period and after weaning and in this way could have opportunities of learning about safety, shelter, grazing grounds and sexual approach. The calf in the summer months preceding the rut may attempt to mount the cow and courting bulls are not tolerated by the cow unless they are tolerant of the calf. The pattern is somewhat different with the Wyoming Elk or Wapiti (*Cervus canadensis*). Altmann (1963) has noted that while the calf tends to be born in relatively secluded places, the cow has some visual, auditory or olfactory reference to the herd. She leaves the calf in seclusion to graze but returns at intervals to nurse. After 18 to 20 days, the calf follows the cow back to the herd where there is evidence of a status structure, the calf following the status of the mother when, for example, shelter from a hailstorm is limited. In the pre-rut period of the autumn, with the disintegration of the herd, the attachment of mother and calf becomes stronger but with the imminence of the new calf in the spring, the cow becomes increasingly hostile to her offspring and will threaten (head high, ears folded back and stamping of front feet) and drive it away. Some bond remains, however, and as the new calf progresses, the hostility to the yearling is relaxed. Should the new calf die, the yearling may be permitted to nurse.

Hafez (1962) has also noted the savage way in which equine mares drive off attendant yearling foals, ensuring in effect fewer intrusions to blur the attachment of a new foal to the mother. It is also of interest to note from Jane van Lawick-Goodall's film (1968) that the chimpanzee mother with an infant in her arms would tolerate intense observation of the infant by a young male chimpanzee but gently pushed away his hand when he appeared about to touch the infant.

The practice, however, does vary with different primates. Among the Langurs (*Presbytis entellus*), Jay (1963) has noted that females gather round the mother as soon as they notice the new-born and several of them may groom the mother. This all takes place without threat or gestures or apparent involvement of dominance-status. Some hours after the birth, the mother even allows other females to hold the infant and carry it away for distances which by the end of the second day may extend to 75 ft. She always remains aware of its position, however, and if there is a disturbance in the group can quickly retrieve it, regardless of the status of the other females. Male langurs appear relatively indifferent to the young. Very different is the behaviour of Baboons (*Papio ursinus*) in which a birth and the newborn appears to engage the attention of the whole group. The mother at first resists all attempts by others to touch it; but after a week or ten days, young juveniles and infants may be allowed to reach over the mother and touch the newborn during a grooming session. Infants are an important factor in the cohesion of a group of baboons. They confer status on the mother. Individuals of all ages tolerate them and even old males have been observed to allow infants to climb over them and to carry infants clinging to their bellies.

Blauvelt's (1955) study with goats, however, included a very interesting finding. Those kids about whom the doe had been prevented from maintaining a 'safe' area did not develop normally. Not only were they backward socially in that they shrank from invitations to play or reacted by aggressive butting; but at the age of six months, they were below normal weight. In several cases, their social reactions were so poor that other kids reacted to them as to an inanimate object, such as a rock or tree stump. Liddell's (1960) study also emphasised the long-range effects of early separation. Kids separated from the does for periods of 30 minutes to 2 hours, soon after birth, withstood the strain of his conditioning experiments far worse than controls. Within the first year of their life, 19 out of 22 of this group had died compared with 2 out of 22 of the controls. One surviving female, which had been separated from her own mother soon after birth, was found to be a most inefficient mother to a male kid who proved to be indifferent to the females of the flock.

A further aspect of early attachment has been demonstrated by Harlow and Zimmerman (1959) and Harlow and Harlow (1961). Infant Rhesus Monkeys, separated from their mothers at 6 to 12 hours were exposed to two types of 'mother', namely, crude wire-frame 'mothers' fitted with nipples giving milk and the same type of frame covered with thick towelling. Others were bottle-fed and reared in individual cages which permitted them to see and hear other monkeys, but had no bodily contact. There could be no doubt of the preference and long-lasting attachment of the monkeys for the surrogate mother covered with soft, yielding cloth. Removal of the cloth was always associated with deep emotional stress and the young monkeys, when disturbed, always fled to the clothed model. However, while originally it was hoped that these animals and those reared in individual cages would provide a breeding stock, both groups proved to be worthless in this respect. The reason is apparently that they lacked opportunities for sexual play, which in normal circumstances is fragmentary at first but begins to reveal sexual differences by about the end of the first month, with the males increasingly adopting more aggressive postures and attempted mountings, while the females show more passive behaviour and patterns of crouching and presenting. From about six months onward, there is little overlap in the play of the different sexes. Normal, social development in these and many other animals would appear to require not only a strong early attachment to afford some basis of orientation for emotional life but also some stable social structure in which many aspects of adult behaviour can be acquired within the moulding influences of the group and the learning associated with the incipient postures and the trial and error features of play.

Harlow and Harlow (1961) were able to study the infant–infant system of interaction further by allowing infant Rhesus Monkeys in groups of four to enter daily for 20 minutes a cage containing a number of play materials and devices. Between the play sessions they lived in cages with cloth-covered, surrogate mothers. The heterosexual activities followed the above stages, the limited opportunities for play apparently being sufficient to permit normal, sexual development. The play itself also followed a sequence of stages. First, the infant monkeys played with the objects in the room but not with each other. This was followed by harmless rough and tumble play and then approach and withdrawal play with the animals chasing each other to and fro, but making little bodily contact. The fourth stage was more integrated and incorporated features of all of the three early stages with occasional bursts of frenzied activity. Towards the end of the first year, a fifth stage of aggressive play appeared which included biting and pulling, but little bodily harm. The aggression gradually increased in intensity until a dominance hierarchy was established.

A group of 55 monkeys deprived of mothering and the companionship of other monkeys for the first three years of life, on the other hand, revealed some highly abnormal behaviour. Observation over six years indicated that for long periods they sat passively in their cages, staring into space and paying little attention to animals in other cages. They showed much stereotyped behaviour such as rocking and clasping the head or body in their arms and patterns of self-punishment, including biting and tearing the flesh of the arms. Their reaction to other monkeys in trial situations was usually anti-social and included much fighting. Over several years, no motherless male ever achieved intromission and only one motherless female was impregnated. In general, the females either attacked or avoided males who approached them. This result is broadly similar to the results with cocks reared in visual isolation in Kruijt's extensive studies (1962, 1964) of the Burmese Red Jungle Fowl (*Gallus gallus spadiceus*). Instead of the customary dignified waltz and courting adapted to the responses of the hen, only fragmented patterns were observed, punctuated by bouts of excessive reaction of escape or aggression, even towards innocuous stimuli, such as a sparrow dropping into the cage, and to the hens who could be viciously attacked and mounted the wrong way round.

Even among animal species there is much that can go wrong and prevent the attainment of a socially adjusted animal competently reproducing offspring who in turn acquire the social skills. Strong innate tendencies to attachment in themselves are not a guarantee of success. If the social experience is not appropriate, bizarre results can follow. Von Frisch (1957) reports the case of a young Purple Heron (*Ardea purpurea*) raised in captivity which addressed its courting and mating behaviour towards its human partner. Strangers were attacked whereas the familiar man was the object of greeting displays and even sexual 'treading' on the shoulders. Schutz (1965, 1966) has shown, too, that Mallard drakes can be rendered homosexual by rearing them in small flocks of five to ten for 75 days or more. When released on the lake at Seewiesen, Bavaria, among a great variety of water fowl, including many female Mallards, they formed homosexual pairs in which neither was prone to accept the passive role of the female. The relationship nevertheless persists for many years and possibly, subject to continued study, for the remainder of the bird's life.

ATTACHMENT IN CHILDREN

Features of attachment and solicitude and consequences in human children provide the basis of such an extensive literature that whole books have been devoted to the field. Inevitably, in studies of delinquency and educational backwardness, the contribution of social practices, relationships within the home and school, in short, all the features deriving from a particular culture and system of communications, are emphasised. But there are other factors and features which begin to operate long before any organised, linguistic communication has developed in the child.

In a sensitive study, Schaffer and Emerson (1964) noted the reactions of 60 infants (31 male) in variety of circumstances such as being set down after being held, after being left alone in a room or in a pram in the street. With variations deriving from family organisation and illness, they found that attachments to specific persons had usually been achieved by about the age of 9 months. The peak of their graph for specific attachment was at 41 to 44 weeks and the object was in 93 % of cases the mother; but an aunt, father or grandmother could be the specific object if they spent more time with the infant. In general, the infant with a strong attachment to a principal object tended to be more afraid of complete strangers on first encounter but, overall, had a greater range of minor attachments.

This general outline of development is broadly consistent with other findings. Bühler and Hetzer (1928) indicated that their infant subjects were between five and seven months before they distinguished between a smiling and scowling face. Kaila (1932) came to the conclusion from a study of 71 infants of the same national background from one institution that between the age of three and six months, the baby's smile in response to a smiling human face is a reaction to a pattern or Gestalt and is not imitative or interpretative. The essential features of the configuration were a smooth forehead, two eyes, a nose and movement in the face such as smiling or nodding the head. When the observer covered his eyes or turned his face in profile, smiling ceased. Spitz (1946) reported similar findings from a study of 251 infants (112 female) from private homes, delivery clinics and nurseries and of white, coloured and North American Indian race. The important elements were a forehead, two eyes, a nose and some movement in the mouth-nose region. If these elements were present in a scowling face, only one of 142 infants tested failed to smile. A crude Hallowe'en mask, tapped from behind in the mouth-nose region, gave the same results. After the age of six months, the efficacy of the masks began to decline and, as Ahrens (1954) and

Ambrose (1960) found, more differentiation is required to meet the increasing learning of the child. After six months, the smile is evoked by faces with which the child is familiar. It is of interest that throughout Spitz's extensive study, children living in an institution, as a class, tended to be retarded in revealing these responses when compared with those from a kindly, home background.

A general finding from many studies is that animals which have incurred sensory and maternal deprivation are, as a class, less resistant to stress and less efficient in a variety of tasks. Clark et al. (1951) found that inbred Scottish Terriers reared in isolation, apart from the unavoidable association with their human attendants, were inferior in simple test situations, more hesitant and below the weight of controls. Similar results were reported by Thompson and Heron (1954). Melzack and Thompson (1956) and Nissen et al. (1951) have drawn attention to the repetitive movements of captive chimpanzees. The pacing to and fro of captive animals might be an attempt to gain more and varied stimulation and, if boredom and stress were reduced, could be self-reinforcing and prolonged. Levy (1944) has reported similar observations from a wide range of captive animals and has noted, too, the greater incidence of stereotyped movements among children in orphanages. Clearly a vast literature and field for research is indicated and space is available to mention only a few field studies; but they will serve to indicate the nature of the problem and one aspect of continuity between animal and human behaviour.

As emphasised by Margaret Ribble (1943) in *The Rights of Infants*, the human neonate has an uncertain hold on life and in the early period, frequent and gentle bodily contact with another person, gentle massage and rocking with low rhythmic sounds have a stabilising effect on many functions, including breathing. Successful sucking responses often depend upon a sympathetic approach and deft stimulation of tongue and palate. The old-fashioned nannie gently rocking and crooning to the child held gently against her body was giving the child a good start early in life. There is ample evidence to show and the pathetic attachment of Harlow's Rhesus Monkeys to the cloth mother is an indirect illustration that a definite early attachment is needed in life. Spitz and Wolf (1945) from a study of 123 infants aged from 6 to 12 months reported in detail on 19 cases of whom they had observed prolonged weeping and later insomnia, poor response to social approaches and poor development. Investigation showed that in all cases, the child had been separated from its mother between the sixth and eighth month. They likened this state of 'anaclitic depression' to a state referred to by Abraham and Freud as pathological mourning for the loved one.

As the child grows older and more aware of its social environment,

the basic sense of attachment, hitherto emphasised by physical contact, is manifested in a need for solicitude, the conviction that some-
one cares about oneself. If the conviction is achieved, there is a
reference point, a basis of orientation and assurance which is an important aid in coping with the doubts and anxieties associated with
growing up, until the developing personality has acquired the width
of experience and confidence to ensure independence. If the conviction is not achieved, there is not only the possibility of bewilderment
and indifferent social response but also the feeling of having been
cheated, of active hostility against family, order and society which
can take such varied forms as delinquency or even failure at school
in order to hurt the parents. Bowlby's monograph, *Maternal Care
and Mental Health* (1951), included a vast amount of data ranging
from deprivation in hospitals and institutions to war-time separation
and hostility. The evidence all points in the same direction; but it
should not be concluded that maternal deprivation renders delinquency or abnormality inevitable. Correlation coefficients only offer
a basis for probabilities. Lack of solicitude increases the probability
of abnormality. It does not ensure that this result will certainly
follow and the possibility is thus open that exceptional cases will
triumph over every hardship and deprivation. A few studies, however, will show how serious the possibilities can be.

   Durfee and Wolf (1933) found that the retardation among 18
children in an institution was most severe among those who had spent
more than eight months of their first year in an institution. Studies
by Goldfarb (1943a to 1945) showed differences in favour of children
who had gone direct from their mothers to foster homes where they
remained, when compared with other children in foster homes who
had spent their early years in institutions. Bowlby et al. (1956) reported on a group of 64 children (41 boys) who had been admitted to
a TB sanatorium before the age of four years between January 1940
and November 1948. In 1950, when the children had left the sanatorium to be scattered among 58 different schools and the ages
ranged from 6 years 10 months to 13 years 7 months, several controls
were made with three children of the same age, in the same class and
taught by the same teacher. The mean IQ of the sanatorium children
was three points lower than that of the controls, a difference which
approached but did not attain statistical significance. However, it
was noted that the sanatorium children were poorer in concentration,
more prone to day-dreaming, more indifferent to competition in
school work and more prone to become violent in play. Humphrey
and Ounsted (1963) in a study of 80 legal adoptions (53 boys) in
Oxford noted the interesting fact of the greater number of children
adopted before the age of six months who attain an IQ of 110 or

better, when compared with adoptions at a later age. The contribution of emotional factors to school work and performance on intelligence tests is probably considerable and, as yet, not completely understood; but it could be that those adopted in the first six months of life have a better chance of forming a strong attachment to the foster parents. Other possible variables are that the helplessness of the younger infant calls forth a greater emotional attachment from foster parents and also that those foster parents who are prepared to adopt young infants are, as a class, more dedicated and efficient foster parents. Another observation from the study was that children adopted after the age of six months were more apt to steal or destroy property.

A frequent source of emotional stress to children is the severance from home and family associated with a period in hospital. Prugh et al. (1953) studied two groups of children, each of 100 children, ranging in age from 2 to 12 years, closely matched for age, sex and type of illness of which there was a wide range. The average stay in hospital for the control group was eight days and for the experimental group, six days. The experimental group received daily visits from parents with whom the hospital staff cooperated. There could be no doubt from the results that hospitalisation was less disturbing to the experimental group. The other interesting result was the difference in the incidence of severe reactions with age—50 % among controls, 37 % in the experimental group for those under three years and 30 % and 7 % for those from 6 to 12 years. Schaffer (1958) has also indicated an age variable in a study of 76 infants admitted to a children's hospital between the ages of 3 to 51 weeks with a stay ranging from 4 to 49 days (mean 15·4 days). Two syndromes were observed on returning home. The 'global' syndrome was more typical of those under seven months. The child seemed engrossed in the environment. Toys presented to him would be disregarded while he scanned the room and mothers and visitors had the unpleasant feeling that the child was looking through them. The 'over-dependent' syndrome was more commonly found among those returning after seven months of age and was characterised by excessive crying. One possible explanation is that the younger child has not yet formed entirely clear attachments to specific objects and is more merged with his environment which he is slowly coming to know. After the rupture in his experience, he is trying in the global syndrome to establish again his orientation. The older child is more aware of himself as an independent being and has gone further towards establishing attachments to specific objects in his environment. After the rupture, he feels more bereft and seeks and clings more violently to the mother, the principal object of his former attachments.

Turning to later years of childhood and to adolescence, the many studies of delinquency provide a consistent story. Stress or anxiety in the home or a broken home is associated with a greater probability of delinquency and crime. This has been the consistent finding from studies by the Gluecks (1934), Menut (1943), Bowlby (1944), Stott (1950) and many others. It is noteworthy, too, that the League of Nations survey (1938) of 530 prostitutes revealed a high incidence of disturbed and unsatisfactory early life and it may be that failure to form effective attachments at this time reduces the probability of forming a satisfactory and lasting emotional attachment later on.

The foregoing brief survey will have indicated some of the very effective forms of attachment of mother and young which have appeared in the course of evolutionary development and some of the consequences when, for various reasons, the processes of attachment are damaged or prevented. Some indirect indication of the importance of attachment is provided in the phenomenon of imprinting. As indicated by Smith (1962), a variety of intermittent stimuli will attract domestic chicks and they soon acquire the ability to discriminate between the familiar and unfamiliar stimuli. Movement across the frontal, parallel plane, movement directly ahead in the sagittal plane, the presence of contrary movement, the presence of colour and simultaneous presentation of visual and auditory stimulation are all conducive to approach and following by the new-born chick. In fact, intermittent visual or auditory stimuli alone are sufficient to attract the young bird. The hen and her normal movements, however, embody very much more than the minimum stimulation necessary to attract the chick. There is movement in both sagittal and frontal, parallel planes, contrary movement of legs and feathers of different colours and brightness, and simultaneous presentation of sound. Viewed as a stimulus combination, the hen manifests a considerable degree of perceptual redundancy. Moreover, those chicks which follow have an increased chance of finding food from the hen's more powerful scratching and the possibility of further reinforcements from the eventual warmth and protection of the hen's body. Viewed in another way, on a rough, engineering analogy, it is a 'fail-safe' arrangement and illustrates in the perceptual field a trend apparent elsewhere, namely, that frequently when an issue of survival may be involved, evolutionary development has resulted in more than the minimum provision. Human beings, for example, have more than the minimum lung and kidney tissue necessary for survival, but the additional tissue is in one sense an insurance against exceptional circumstances.

It will be apparent, too, that in many species the period of attachment of the young is necessary not only for their survival but also for the development of full social competence. The studies of mammalian young, including monkeys, apes and human beings in particular, indicate that the attachment itself must be supported by experience within society and the trial and error experiences of play. The emergence of play groups relatively independent of the adults occurs with sufficient frequency to suggest that it has biological significance. Possibly the awareness of physical and social inadequacy is more acceptable and less disturbing if the individual is aware of others of similar standing and by observation is aware of progression to and from his present position. In human communities, there are opportunities for remedial steps for lack of attachment in the young; but it is possible that the demands for social competence are greater for the developing child and adolescent and that they may be more aware of the discrepancies between their attainment and the expectations of society than is possible in other communities of animals. Certain it is that the period of dependence of the young is greatest in the human species and the long period of association with the family is not only potentially more important, but has evolved like any other feature of animal behaviour and, as such, is worthy of study for its biological significance. Against this background, the intrusion of any features of housing and working conditions which lessen the influence of the family are also worthy of consideration.

4

FINDING THE WAY ABOUT

Human observers, impressed by the difficulty and aware of the many
conscious influences from observation which they use in finding their
way about, are apt to be impressed by the sheer scale and accuracy of
many animal migrations, and feats of navigation. The Arctic Terns
(*Sterna macrura*) nest far north in the Arctic within a thousand miles
of the pole and fly south, down the west coast of Africa, to winter on
the Antarctic pack ice. There are long stretches between the southern
tip of Greenland and the West African coastline where there are no
guiding land masses. The Wilson's Petrels (*Oceanites oceanicus*) fly
from their breeding grounds in the islands of the Antarctic, up the
Atlantic coast of South and North America and, while they do not
penetrate as far north as the Arctic Terns, a journey of at least 7,000
miles is regularly achieved. The Bristle-thighed Curlew (*Numenius
tahitiensis*) nest in Alaska but winter in the islands of the Polynesian
group about 6,000 miles away. The shortest sea crossing is 2,000
miles, without landmarks, and on the return journey, Alaska is but
one small point, subtending an arc of 45° stretching across the
islands. Even more impressive is the performance of the Great
Shearwaters (*Procellaria gravis*) which range the whole Atlantic as far
north as 60° but return to breed on the islands of the small, isolated
Tristan da Cunha group. Manx Shearwaters (*Procellaria puffinus*)
have made their way back to Skokholm from Venice in fourteen days
(Lockley, 1942) and one from Boston, USA, in twelve and a half
days (Matthews, 1953). Both were journeys outside of the Shearwater's normal direction of activities and were accomplished too
quickly to admit complete reliance on trial and error. Moreover, the

Atlantic journey was made over the sea with no landmarks and must have included some time for resting and feeding.

It must also be remembered, however, that many animals are not called upon to perform such feats of navigation in order to complete their life cycle. Their normal range is provided with many clues for orientation or guidance. The domestic dog is a familiar example. By urinating at points to which it is attracted by the odours of other dogs, the individual not only has a basis for orientation but also a means of acquiring information about the presence of other dogs in the area, their physical and breeding condition and their numbers. Seton (1910) noted that the area ranged by wolves in North America is liberally provided with such reference points. If in addition, the scents from the trail are used, it is possible to conceive of individual animals moving about within a matrix of references giving spatial orientation and an immense amount of social information which can determine its movements and the spacing between individuals. So too the practice of the American Black Bear (*Ursus americanus*) of standing at full height and gnawing and salivating on small trees could serve a similar purpose, giving information as to the recency of the visit, the size of the animal, its power, the condition of the teeth and age.

Domestic cattle when free to range widely very often adopt regular camp sites where they ruminate and between which they establish regular trails. Regular dunging places of many wild mammalian species are conceivably not only helpful in orientation but also afford information about recency of use, the age and condition of the users and numbers. Even animals separated from the herd could not move far without encountering information in the form of scent trails, well-worn visible trails and rubbing places so that it would be difficult to conceive of large numbers of animals which move over the earth's surface being completely disorientated or 'lost' in their normal environment.

THE PERFORMANCE OF BEES

The possibilities of more complicated forms of navigation definitely arise from consideration of the remarkable work over many years with bees by von Frisch (1967). Survival is not easy for bees in temperate latitudes. They have a limited season in which to gather supplies of pollen and nectar to sustain them over the winter months and perforce must work efficiently in the limited time available. The evidence is that they can not only navigate, but also communicate the position of their finds to other bees. If a good source is found within 50 yards of the hive, the 'round' dance is performed by the finder. Other bees join the dance and the message is that a good source of food corresponding to the odour on the finder's body is available if the dancers will search around the hive within a radius of about 50 metres. It was noted that, for finds at 50 metres, some 'waggling' was becoming evident in the dance and, for finds beyond 100 metres, the 'waggle' dance was employed, so called because the dance consists of two semicircular loops alternatively on the left and right side, resembling a figure of eight and on the straight central section, the tail is wagged. As the distance of the find increases, the tempo of the dance declines. For 500 metres, 6 circuits per 15 seconds was the average and for 10,000 metres, slightly less than two (Fig. 2).

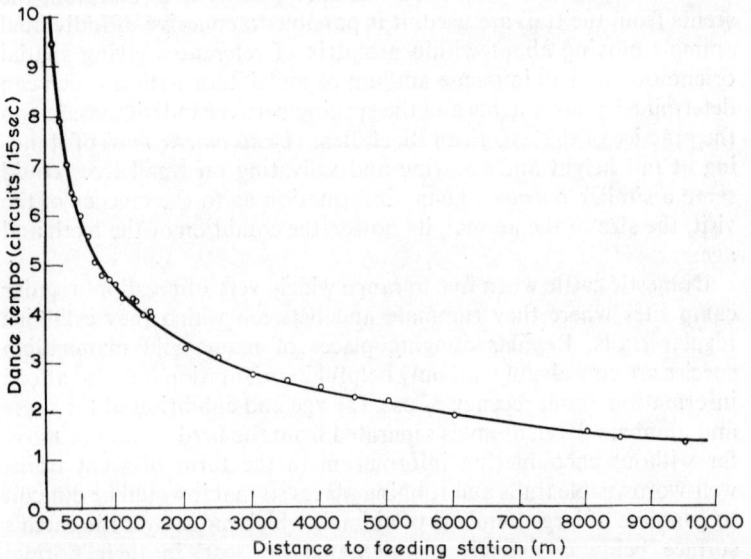

Fig. 2

Graph showing the relationship between the tempo of the bees' dance and the distances involved. The curve is based upon average values from experiments of von Frisch, Heran, Knaffl and Lindauer and involves 6267 dances. (After von Frisch, K., *The Dance Language and Orientation of Bees*, Oxford Univ. Press and Harvard Univ. Press, 1967, originally published as *Tanzsprache und Orientierung der Bienen*, Berlin–Heidelberg–New York, Springer, 1965)

If the dance is performed on a horizontal surface, the finder orientates the straight central section of the figure of eight in the direction of the find, maintaining the same angle to the sun's position as on the previous flight (Fig. 3). If the waggle dance is performed on a vertical surface, such as the surface of the comb, the sun's position is

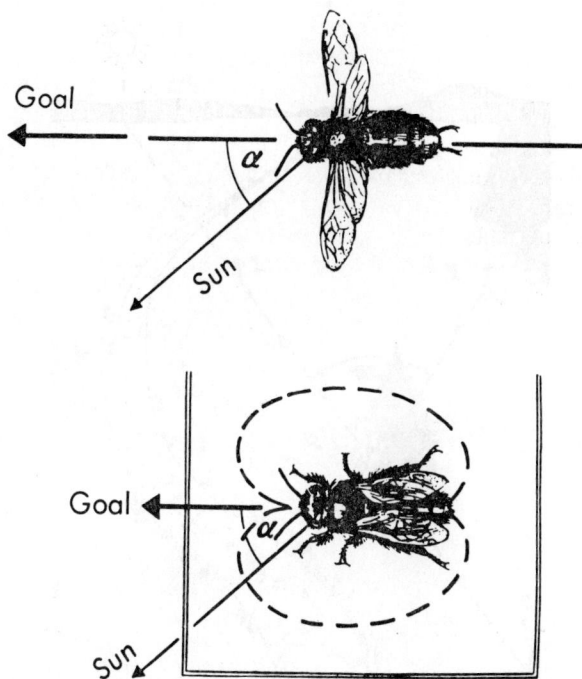

Fig. 3

The method of indicating direction on a horizontal surface. During the waggling run up the middle of the figure of eight dance (lower figure), the bee takes a direction so that she views the sun at the same angle (α) as during her previous flight to the 'find' or feeding place. (After von Frisch, K., *The Dance Language and Orientation of Bees*)

represented by the vertical line of gravity and variations to left or right of the sun's position are represented by corresponding angular variations from the vertical (Fig. 4).

Bees can orientate not only with reference to the sun itself, but also with reference to the polarised light, which depends upon the position of the sun, provided that some blue sky is visible. Under severely overcast conditions, the bees cannot orientate during the waggling run. These methods communicate the general direction but there is ample evidence that individual bees use local, visible landmarks. In one of von Frisch's experiments the feeding station to which the bees had been flying for half an hour was situated one metre from a small field sign. When, on the following morning, the station was moved about 10 metres, it was noted that the bees flew to the signpost and then flew about it before making their way to the

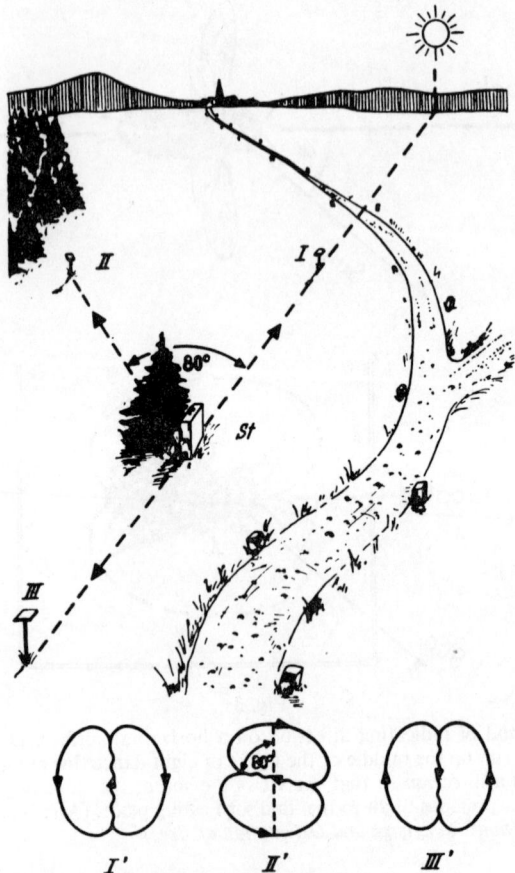

Fig. 4

Summary of the bee's method of indicating the direction of the 'find' by tail-wagging dances on the vertical comb. St is the beehive, I', II' and III' are the plans of the dance to indicate directions from the hive of the feeding stations, I, II and III, respectively. (After von Frisch, K , *The Dance Language and Orientation of Bees*)

feeding station. Another example of the use of intermediate, visual goals is the persistent way in which the bees at times adhere to them. On one occasion during August, von Frisch designed an experiment to discover if the bees' 'language' or system of communication included a means of indicating elevation. Near Brunnwinkl, his base of operations on the Wolfgangsee in Bavaria, stands a steep cliff, the Schafberg, 1,780 metres high. From 15 August to 19 August, the observation hive stood at the base of the cliff and was then taken back to Brunnwinkl where the bees flew around for a week before the hive was again taken to the Schafberg; but this time it was placed on top of the cliff. When the hive entrance was opened, von Frisch was at the former location of the hive, setting up a feeding station. Almost immediately, some eight or ten bees arrived; but only the base upon which the hive had formerly stood was available. The bees did not go to this exact spot but hovered persistently about two metres to one side of it, over a prominent rock and on that side of the rock which faced the former position of the hive. These bees in orientating flights and in their rush to return to the hive had not registered the new position above the old one and apparently were hovering around the rock, the nearest intermediate goal to the former position. It appears that the bees use landmarks for their *own*, individual orientation and, as this and other experiments by von Frisch indicated, the language of the bees has no 'words' or way of indicating the landmarks or the features of elevation.

It is of interest to note in this context that the digger wasp (*Ammophila campestris*) also makes use of visible landmarks in a comparable way. Suspecting that pine trees projecting above the heath afforded a means of orientation to their holes, Baerends (1941) introduced dwarf pines in pots into the landscape and was able to control the movement of the wasps at will. For bees, the distance of the landmark from the objective is apparently important. In an experiment carried out in July 1953, von Frisch trained marked bees to fly 180 metres to a feeding station which lay 2° east of south of their hive (Fig. 5). The flight path, however, lay along a line 5 metres from the edge of a wood. The hive was then moved to another field and the feeding station placed at the same distance and again in the direction 2° east of south; but this time the direct line from hive to feeding station was away from and almost at right angles to the wood. Of 35 foraging bees marked on the preceding day, 22 were recovered. Only 5 had flown in the correct southerly direction, 16 had followed the line of the wood which in this case was 5° south of west from the hive and one had flown along the line of the wood in the opposite direction. When the distance of the bee colony from the margin of the wood was increased, first to 60 then to 200 metres, the bees were found to rely less and less on the line of the woods and more on their use of the 'sun compass'.

Now the sun as a basis for orientation or as a compass is not very helpful unless allowance is made for the time of the day. A measure of orientation is no doubt afforded to birds migrating over long distances by the awareness that the east–west arc of the sun is 'over there', i.e. north or south of their current position. Over the relatively short distances travelled by foraging bees—10 kilometres would be a very

C

long flight—this information would be of no value. Experimental evidence indicates that bees have much more precise and effective methods of finding their way about.

In one of his innumerable investigations, von Frisch used a colony of marked bees foraging from a hive in Brunnwinkl. A feeding station was set up 200 metres to the west of the hive (see Fig. 6). The flight path between hive and station lay between houses and trees and across a small bay of the Wolfgangsee. The bees had been making this flight for several days. Early in the morning of 24 September 1949 and before the bees were flying, the hive was moved to a region, unknown to the bees, five kilometres from Brunnwinkl and set up in the centre of a broad, level field. Four feeding stations, each affording the now familiar diet (sugared water with two drops of oil of anise) were set up, each 200 metres north, south, east and west of the hive. At each station was an observer with instructions to kill the bees as they arrived, so as to prevent communication between them. At 8.25 a.m. the hive was opened. Orientation flights began at once and indicated that the bees were reacting to the altered situation. In all, of 29 marked bees which had visited the feeding station to the west of Brunnwinkl on the previous evening, 27 came to a feeding station, 20 to the west of the hive, 5 to the south and one each to the north and east; but during the morning, the sun stood in the south-east and for the flights of the previous afternoon, the sun was in the south-west. Von Frisch feels that the removal of the hive by man to a new location is an 'unbiological' event. After all, in natural circumstances, colonies of bees migrate and select their new place of abode. For this reason, they would tend to seek the feeding station where they had become accustomed to find it, to the west of the hive. He also offers an explanation for the five bees who went to the southern feeding station. The southerly wind on the day of the experiment could have borne the scent towards the searching bees. Again, young bees have to learn the course of the sun. Until this is known, they tend to follow a path at the same angle with the sun as on previous successful flights. During the last flights on the previous afternoon, with the sun in the south-west, the feeding station west of the hive was about 40° to the right of the sun. On the next morning, with the sun in the south-east, the southern feeding station was about 40° to the right of the sun's position.

That bees require a little time and experience to learn the sun's course and to extrapolate from its different positions has been demonstrated by Lindauer (1959). Bees were developed in an incubator and allowed to form a colony in a cellar where they had no view of the sky for several weeks. On the first day of free flight, they were trained to fly to a feeding station in a particular compass

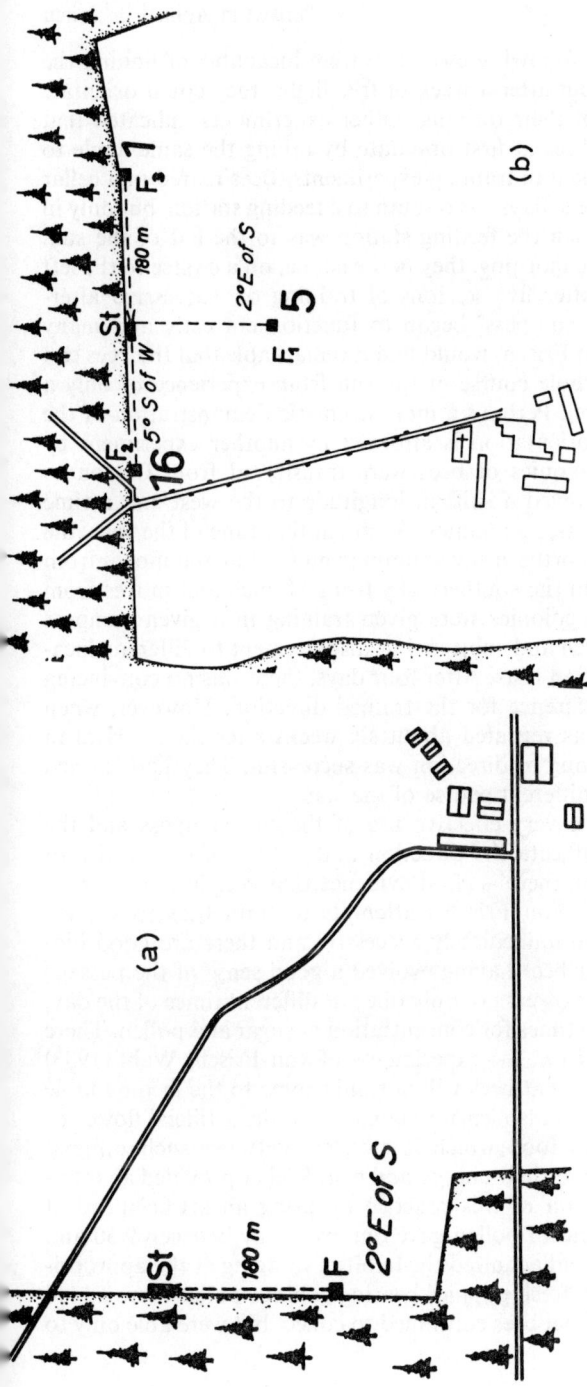

Fig. 5

An experiment indicating that bees are, for a time, influenced by local features in finding their way about, although they have no way of communicating topographical features to other bees. In (a), St is the hive. F is the feeding station in training flights. The flight path from St to F was 5 metres from the edge of the wood in direction 2°E of S. In (b), the situation on the following morning is shown. The hive is 5 metres from the edge of a wood. Feeding station F_1 is 2°E of S from the hive. F_2 and F_3 are new feeding stations, adjoining the wood. Of 35 foraging bees marked in the training flights on the previous afternoon, 22 returned to feeding stations, 5 only to the correct one F_1. In the short term, the local features had proved more influential than the sun compass. (After von Frisch, K., *The Dance Language and Orientation of Bees*) direction. On the following day, they were incapable of finding the direction again; but after a week of free flight, they could orientate in the direction of their training. Other experiments indicated that the cellar-reared bees at first orientate by taking the same angle to the sun as that taken in training experiments. Bees reared in a cellar were trained for three days to go south to a feeding station, but only in the afternoons when the feeding station was to the left of the sun. When tested in the morning, they flew east, i.e. on a course to the left of the sun. But after five sessions of training on successive afternoons, their 'sun compass' began to function correctly and many observers, like von Frisch, would find it remarkable that the bees can extrapolate the whole course of the sun from experience of only a section of its travel. Perhaps a more dramatic demonstration of the bees' powers of orientation is afforded by another experiment by Lindauer. Two colonies of bees were transferred from Ceylon to Munich. This involved a shift in longitude to the west and a time difference of 4 hours, 36 minutes. Again, at that time of the year, the sun is seen in the northern sky at noon from Ceylon and moves from right to left, but in the southern sky from Munich and moves from left to right. The colonies were given training in a given compass direction at Munich and subjected to displacement to different locations and orientation tests. After four days, there was no convincing evidence of a preference for the trained direction. However, when the experiment was repeated about six weeks after their arrival in Munich, the training to direction was successful. They had learned to orientate to a different course of the sun.

In addition to a very effective use of the sun compass and the ability to communicate the direction and nature of discoveries of food to other bees, there is good evidence that bees have a form of 'biological clock'. Von Frisch's attempts to train bees to specific feeding times were immediately successful and there are good biological reasons for bees having evolved a good sense of the passage of time. Different flowers not only open at different times of the day, but have optimum times for concentration of sugar and pollen. There is good evidence from the experiments of von Frisch, Wahl (1933) and Kleber (1935), that bees will not only come to the feeding table at specific times, but also learn to come when the artificial flower or feeding dish offers food which is sweeter. Between such optimal times, they conserve their energy and rest. Kleber provided an interesting demonstration of this practice by using an artificial bed of poppies whose yield of pollen gave out every day between 9.30 and 10 a.m. The bees had acquired the habit of foraging at the appropriate time. One day, fresh poppies bearing copious pollen were added to the bed. Individual bees continued to collect but were able only to

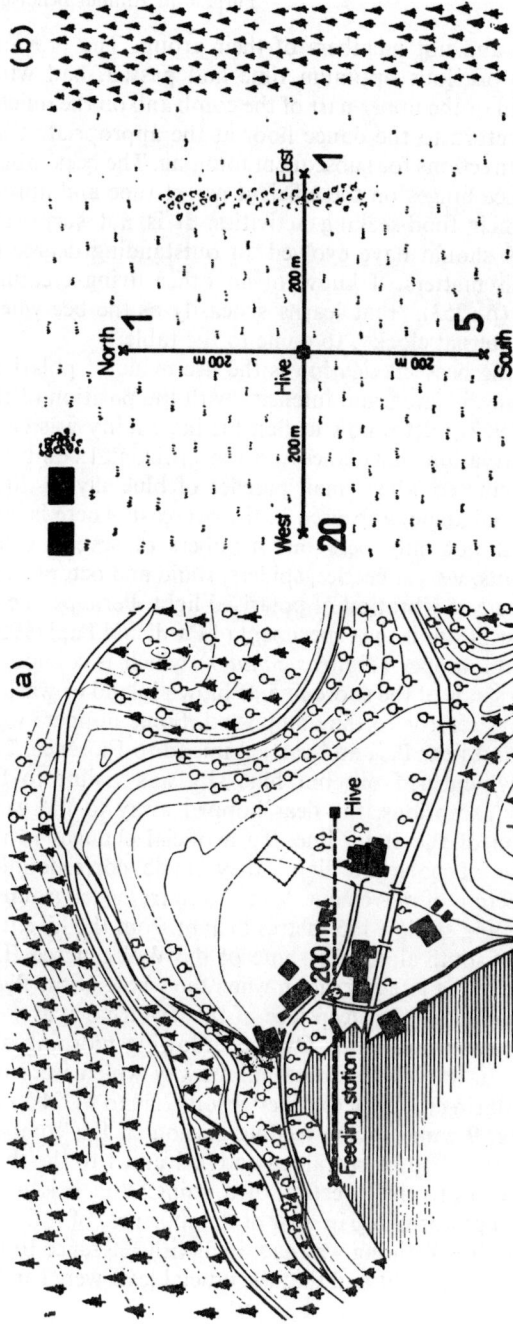

Fig. 6

(a) shows the path in training flights between hive and feeding station. It passes over many topo-graphical features; houses, trees and a bay in the lake. (b) shows the plan of the crucial experiment. The hive was moved to an entirely unfamiliar environment and four feeding stations set up at the cardinal points, each 200 metres from the hive. Of 29 trained bees, 27 came to a feeding station with the distribution as shown. (After von Frisch, K., *The Dance Language and Orientation of Bees*) recruit newcomers and not members of their group. The previous foragers, once the previous optimum time had passed, had withdrawn to the margin or the upper part of the comb and on the quieter side. They would return to the dance floor at the appropriate time and there acquire directions for subsequent foraging. The bees' whole struggle for existence hinges on the efficient use of time and appropriate timing of their food-seeking activities. It is not surprising therefore that they should have evolved an outstanding degree of adjustment in such matters. 'I know of no other living creature' wrote von Frisch (p. 253), 'that learns so easily as the bee when, according to its "internal clock", to come to the table.'

Another aid to the bees' navigation is the use made of polarised light which varies in direction and intensity with the position of the sun. Von Frisch (1967) relates how at first his incredulity was challenged by the observation that dances on the horizontal comb became correctly orientated when small patches of blue sky (as little as 10° to 15° of visual angle) appeared in the overcast. There is now good evidence that not only bees but members of several other species including ants, wasps, beetles, spiders, squid and octopus can orientate to the plane of vibration of polarised light. Perhaps one of the neatest demonstrations is the experiment of Pardi and Papi (1952) with the small Sand-hopper (*Talitrus saltator*) which is commonly found in the damp sands of the shore line. If blown inland or washed out to sea, it always returns to the damp sand. Some hoppers were captured on the coast near Pisa and carried inland to Florence. The sun was effectively screened out but blue sky was visible to the animals. With few exceptions, the fleas hopped away to the west, i.e. in the direction of the shore line. To artificial changes in the plane of vibration of polarised light, they made corresponding changes in the direction of movement. Von Frisch trained a group of bees to go to a feeding station 175 metres distant from the hive in a direction 38° east of south along the shore of the Wolfgangsee. The training took place in the late afternoon when the whole of the flight path was in the shadow of the steep face of the Zwölfhorn, as were all preliminary, orientation flights. On the following morning, the colony was set up in level fields some miles away which were unfamiliar to it and, during the time of observation (7.54 to 11.54 a.m.), of 30 marked bees, 9 came to the feeding station in the compass direction to which they had been trained and 0, 1 and 1 to the three stations at right angles to the direction of training. Von Frisch felt that this was a good performance in view of the shortness of training. It will be noted that the training was acquired with reference to the pattern of polarisation of light and that the crucial tests were carried out in sunlight.

All of the displacement experiments and the method of communication by dancing must involve some method of retention or 'memory'. Free-flying bees have also on occasion been observed to perform directed dances when it would have been impossible for them to have seen the sun on their last flights. Even more impressive is the observation reported by Lindauer (1960) of an experiment by Bräuninger. Between 24 October and 1 November, marked bees were fed at a station 385 metres distant from the hive. The weather then became very cold and the hive was moved indoors to a warm room. At 10.5 a.m. on 8 December sugared water was supplied inside the hive. This sometimes acts as a stimulus to dancing. On this occasion, two bees did in fact begin to dance on the vertical comb and, without view of the sun or the sky, gave the correct orientation for their last outdoor feeding place. Bees have on more than one occasion been observed to dance at night, giving fairly accurately the bearing of their goal on their last flight.

Other abilities which aid the bees include olfactory sensitivity which is about equal to that of man but is especially good in detecting the odours of flowers. They react very poorly to airborne sound but do react to vibrations of comb and hive produced by sound. They can learn differences between visual patterns if they are heavily or distinctively indented. Their colour sense is much stronger towards the yellow, green, blue and ultra-violet end of the spectrum and colour appears to be more important at a distance while odours are more influential at close range. All the evidence indicates that the remarkable powers of orientation of the bees owe something to maturation and learning. Young bees are usually occupied with housekeeping duties. After some eight or ten days, they make orientation flights about the hive and the return appears to be aided by olfactory cues and nearby visual landmarks. Bees which have never left the hive rarely find their way home if liberated more than 50 metres from it. After the first orientation flight, there is a very noticeable improvement. Some bees are capable of finding their way back from distances of several hundred metres and these distances can be extended to several thousand metres with further experience of the terrain. The impressive feature is that the bees acquire the orientation so quickly.

MARINE SPECIES

The migrations of marine species form the basis of an extensive literature which, because of the difficulties of observation, is at times incomplete at crucial points. Extensive and regular movements are nevertheless known among many fish of economic value from a number of methods, such as tagging, netting, echo-location and, in favourable conditions, observation from aircraft. Cod (*Gadus morhua*), of which 2·7 million tons were caught in 1965, consists in several stocks covering such diverse areas as the Atlantic coast of North America from North Carolina to Baffin Bay in Greenland, east Greenland, Iceland, Faroes, Jan Mayen Island, the Irish Sea, the North Sea, the North Cape, the Baltic and the Gulfs of Bothnia and Finland. Each has its separate pattern of migration, the general feature of which is a movement to warmer water for spawning which takes place from February to May and, in the Newfoundland area, as late as June. Some of the cod maturing in Greenland waters migrate to the coast of Iceland to spawn, possibly in association with the warm Irminger current. Cod from the Barents Sea come southward to spawn off the Lofoten Islands of northern Norway, in each case a journey of several hundred miles. The familiar migration of the Atlantic Herring (*Clupea harengus*) towards the North Sea coast of Britain commences in the north and becomes progressively later towards the south. This, in former times, set in train a considerable human migration of workers who worked their way down the coast with the fishing fleet to gut and salt the herring. Tunny (*Thunnus thynnus*) spawn in the Mediterranean and the eastern Atlantic off the north-west coast of Africa and journey around the north of Scotland to spend the summer in northern areas of the north sea. They return by the same route. As with many species of birds, older members have been observed to arrive on the summer grounds before the younger Tunny, possibly because they are stronger swimmers and are more experienced in reacting to clues from current strengths, water content and temperatures and may include individuals who have done the journey before. For some or all of these reasons, the older fish, as older birds, would tend to wander less.

One of the most extensive journeys undertaken by any marine species is that credited to the European Eel (*Anguilla anguilla*) which is spawned towards the northern fringe of the Sargasso Sea in an area centred around 26°N, 55°W. Over the next two and a half years, the larvae (leptocephali) grow to lengths of 70 to 80 mm in flat, leaf-like forms which drift in the surface waters of the north Atlantic and are carried eastwards by the prevailing currents and may enter the Mediterranean as unmetamorphosed larvae when between one and a half and two years old. According to Schmidt (1932) the change from the flat, leaf-like form of the leptocephalus to the worm-like form of the elver begins when the organism is about two and a half years old. The final stage is the pigmented, bottom-living elver which is preceded by the stage known as the pelagic 'glass' eel and it is in this form that many young eels arrive off the coasts of western Europe. Normally they arrive off the south-west of Ireland, Portugal, Spain, the Basque coast, Italy and the Nile from October to December. The well-known 'run' up the Severn river begins in February. By March they have reached the Netherlands and Greece and the ascent of Danish rivers begins in April.

Having ascended the fresh water rivers of western Europe, the eels remain there for several years—D'Ancona (1960) mentions reports ranging from 7 to 18 years—and eventually undergo a series of changes associated with the onset of sexual maturity, the most noticeable of which are the changes from the yellow colour of the freshwater eel to the silver of the seagoing eel and the enlargement of the eyes with retinal pigmentation more akin to that of deep-sea fish. At this point, a basic point of controversy enters.

It might be supposed that the mature European eels re-cross the Atlantic, possibly swimming against the general influences of the Gulf Stream until, at a certain area and depth, the appropriate conditions, including those of temperature, are attained and spawning takes place. As far as is known, very few and certainly no reliable captures of numbers of silver eels, which would indicate a trend, have been made at sea. Schmidt's estimate of the position of spawning was based upon the ingenious idea of, in effect, working backwards along the west to east movement of the leptocephali and netting progressively younger samples. This data he obtained by the cooperation of the Danish Mercantile Marine from ships on their normal voyages, since his own research vessel was unsuitable for work in the western Atlantic.

However, there is another complication. There are two recognised species of Atlantic Eels, the European Eel (*Anguilla anguilla*) with a mean vertebral count of 114·7 (range, 110–19) and the American Eel (*Anguilla rostrata*) with a mean vertebral count of 107·2 (range 103–111), and although there is little overlap, difficulties of classification have arisen (see Harden Jones 1968, p. 69 *et seq.*). On present evidence, both are spawned in the Sargasso Sea, the European Eels to the north, the American Eels some five degrees to the south where the temperatures are higher and the temperature gradients with depth appreciably steeper. It is a matter of conjecture as to whether or not the warmer conditions are responsible for the different vertebral counts, but the American Eels do appear to grow faster and metamorphose at an earlier age than the European Eel. There appears to be no simple hydrographic explanation as to why the American leptocephali do not come to Europe in significant numbers, but there is no evidence for this. Either from genetic and/or environmental factors, the metamorphosis of *Anguilla rostrata* does appear to take place earlier and, while it is pure speculation, it may be that they are sufficiently close to be attracted to the freshwater influences from the American rivers (into which they eventually move) at a time in their life cycle when they are sensitive to these influences and sufficiently mobile.

Another problem arises from the eels in the Azores. Schmidt was in no doubt that they were European Eels, but drift charts of the Atlantic would indicate that the leptocephali must be two years old at this point. One possibility is that they are fast developing European Eels or 'stragglers' from the three-year class. Another is that proximity to land and freshwater influences may accelerate metamorphosis. The controversy is further aggravated by the suggestion from Tucker (1959), that the American and European Eels are not distinct species but eco-phenotypes of *Anguilla anguilla*, the observed differences being produced by the different temperatures and temperature gradients which the larvae encounter after spawning. The different distribution is attributed to surface currents and coincidence. It is thus not necessary for the mature European Eel to return from Europe to the Sargasso for spawning since the 'apparent' species of the European Eel (*Anguilla anguilla*) is reinforced from the basic stock of *Anguilla rostrata*. It is an interesting hypothesis which could be refuted if any evidence of mature European Eels moving westward and far from the continental shelf were to be found, when one of the longest, sustained migrations by a marine species would require an explanation.

Salmon

A migration much honoured in stories is that of the Salmon family, Salmonidae. Several species of the genus *Salmo* which includes the Atlantic Salmon (*Salmo salar*) and the genus *Oncorhynchus*, Pacific Salmon, are anadromous, spending much of their life in the sea; but there are some species which are landlocked and there is a report (Calderwood, 1927) of a species in the South Island of New Zealand, apparently akin to the *Salmo salar*, which does not go down to the sea although there appears to be nothing to prevent this. Those which remain in freshwater, however, do reveal some migratory tendencies in moving into the headwaters of their streams.

The chief issues involved in the migrations of Salmon are associated with the 'Parent' or 'Home' stream hypothesis, that the fish in returning from the sea return to the stream in which they were spawned. If the fish did not move far in the sea from the sub-marine influences of the river system in which they were spawned, or are aided by prevailing currents or sub-marine topography, no great feats of navigation would be involved and return to the parent stream might then depend upon such factors as olfactory or gustatory 'memory' or preference and temperature preference. Again, the significance of the return of small numbers to the parent stream are difficult to assess when the extent of predation or disease in the sea is not accurately known. Some indication of the difficulty of the problem may be conveyed by the broad results of an early experiment by Rich and Holmes (1929). From two tributaries of the Columbia river, Oregon, USA, 174,000 downstream migrants, Chinook Salmon (*Oncorhynchus tshawytscha*), were marked. Only 504 were subsequently recovered, of which 99 were taken at the hatchery from which they were released, 5 from nearby tributaries and the remaining 400 by commercial fishing.

Later experiments by Foerster (1934, 1936) showed that in 1927 and 1928 only 0·88% and 1·34% respectively of marked Sockeye Salmon (*Oncorhynchus nerka*) returned to the control 'gate' where Cultus Lake empties into the Fraser river in British Columbia. In 1930 and 1931 (Foerster, 1937), 104,601 and 365,265 downstream migrants were marked. Of these, 1·76% and 0·87% respectively were recovered at Cultus Lake. Commercial fisheries recovered 1·91% and 2·62% in the two years. The high rate of loss is striking but those not recovered did not appear elsewhere, although some 'straying' to other streams is not unknown from other studies. Another feature noted was the higher rate of loss of marked fish, although preliminary tests indicated that the method, by fin clipping, did not appear to involve any disability for the fish. Allowing for these losses, the best estimate is that about 7% of downstream migrants get back into the Fraser river and that about 900 out of every 1,000 are not accounted for, presumably remaining or destroyed at sea. A similar order of loss was reported by Pritchard (1939) from a study of Pink Salmon (*O. gorbuscha*) in McClinton creek on Graham Island in the Queen Charlotte group off the Pacific north-west coast of Canada; as with the Cultus Lake experiments, the number of unmarked fish greatly exceeded the marked fish recovered; but in 1934, of the marked fish recovered (2·72% of marked downstream migrants), 90% were taken in the parent stream and, among the several possible alternative streams, the proportion of marked fish recovered increased with the proximity to McClinton creek.

Further evidence of a trend towards the parent stream and of the high proportion of fish not accounted for is provided by Taft and Shapalov (1938) from two studies with marked Steelhead Trout (*Salmo gairdneri*). The first study involved Waddell and Scott creeks, two small and adjacent streams which enter the sea five miles apart, to the north-west of Santa Cruz, California. Over the seasons 1933–4 to 1936–7, of 7,507 marked fish released in Waddell creek in previous years, 333 were recovered in Waddell creek and 5 in Scott creek. Of the 25,938 marked fish released in Scott creek in previous years, 374 were recovered in Scott creek and 7 in Waddell creek. A similar experiment was carried out in the headwaters of the Klamath river in northern California. Not all steelhead trout descend to the sea but scale markings distinguish those which do. Of 23,280 trout released in Beaver creek in 1934, 70 were recovered in 1936–7 in Beaver creek and 3 in neighbouring tributaries. Of 12,650 released in Fall creek, 45 were eventually recovered there and none in other streams.

Another approach to the 'Parent' stream hypothesis involves the transplanting of eggs from the stream in which they were laid to another where the eggs hatch and the fish begin their development. Not all of these experiments have been successful; but some studies indicated the influence of the environment. Rich and Holmes (1929) took Chinook Salmon eggs (*O. tshawytscha*) from the Willamette, McKenzie and Santian rivers which are part of the Columbia river system in Oregon, USA and transferred them to Herman creek, a small tributary running directly into the main Columbia river. From these eggs, 85,000 marked fish were released from which 51 adult fish were recovered, 4 from the adopted stream and none from the stream in which the eggs were laid. However, it was noted that whereas the normal time for the return 'run' of fish to Herman creek is in the autumn, the fish captured returned in the spring, which is the normal time for the rivers from which the eggs were taken. A similar result has been reported by R. W. Wright (1954) with Sockeye Salmon (*O. nerka*). Those marked fish which were recovered tended to return to the stream where they had grown up, which Wright likened to imprinting. The timing of the 'run', however, corresponded to the timing of the run in the river from which they were taken. The extent of the element of 'imprinting' in determining the return of the salmon would require further careful investigation; but Harden Jones (1968) records that successful runs of Pacific Salmon have been established in Chile, New Zealand (South Island) and the Kola Peninsula (Soviet Union). In all cases, the salinity and temperature of the waters encountered correspond to those of the native habitat.

Several experiments suggest that olfactory cues may be very important in several species of fish. Hasler and Wisby (1951) have shown that Bluntnose Minnows (*Hyborhynchus notatus*) can be trained to discriminate between the waters of two Wisconsin creeks. When the olfactory organs were destroyed, discrimination was not achieved. Volatile organic compounds may be important cues. Again, Wisby and Hasler (1954) captured sexually mature Coho Salmon (*O. kisutch*) migrating upstream in two branches of the Issaquah river, Washington, USA. Approximately half of the fish had their nasal sacs plugged before being released below the confluence of the two branches. A significant majority of recaptured, normal fish selected again the stream of their first choice, whereas more of the treated fish changed streams. It might be conjectured that 'sun-compass' orientation is less needed by salmon in view of their habitat, although Johnson and Groot (1963) have suggested that Sockeye Salmon smolts make some use of celestial cues in finding the outlet to Babine Lake in British Columbia; Groot (1965), in experiments in tanks, has produced interesting if debatable evidence of such orientation.

Hasler et al. (1958) have studied the movements of the White Bass (*Roccus chrysops*) in Lake Mendota, Wisconsin. Fish were captured on their two spawning grounds and released in the centre of the lake about 2·4 km away. Their movements were traced by means of a float on a fine, nylon line. Use was also made of eye caps with some fish. There was evidence of a generally northward movement on clear days which would bring the fish back to the spawning areas. There were difficulties about the degree of correction for drift which gives some grounds for doubt. Rather better evidence comes from the study of the Parrot Fish (*Scarus guacamaia* and *S. coelestinus*) from Bermuda reported by Winn et al. (1964). These fish lodge at night in off-shore caves and return to the shore line during the day to feed. Here they were captured and released elsewhere. Of 62 fish released in bright sunlight, the great majority moved in a direction between south and east with the mode at 135°. Tracking was made possible by line and float which could be fitted with a small light at night. Those fitted with opaque, plastic eye caps or released at night in conditions of cloud cover did not reveal the same consistency of direction; but 5 out of 7 fish that had been kept under artificial day-night sequence, which delayed the photoperiod of the fish by 6 hours with reference to local time, moved in a predicted north, northwesterly direction. This on average represented a change in direction of about 190° clockwise which is more than the sun moved in azimuth during the period of the experiment; but this order of change together with the disorientation at night or in heavy overcast suggests that there is probably some reference to the sun's position.

Clearly there is much work to be done and improvement will be necessary in methods of tracking fish so that the large number not accounted for in so many studies can be reduced. In this way, greater statistical precision could be attained and trends more rigorously assessed. It is conceivable that fish may use a plurality of cues and that species differ in the combination of cues used; but be that as it may, sufficient evidence has been mentioned in this short study to indicate that many species of fish have ample cues and capacity of response to them to orientate to advantage within their environment. The statistical evidence would make the case for purposive behaviour over long journeys more convincing.

JOURNEYS BY BIRDS

The considerable distances covered by such migrating birds as the Arctic Tern, Wilson's Petrel, the Bristle-thighed Curlew and Great Shearwater in the course of their regular migrations have already been mentioned and so, too, the 3,050-mile experimentally induced journey over open ocean of a Manx Shearwater from Boston to Skokholm. To these might be added many others: the return of Leach's Petrels (*Oceanodroma leucorrhoa*) from the coast of Sussex, England, to an island off the coast of Maine, USA, two of them in less than a fortnight (Billings, 1968), and perhaps the longest trans-ocean homing flight for experimental purposes (Kenyon and Rice, 1958), the return to Midway Island in the Pacific from the Philippines, a distance of approximately 4,000 miles, by a Laysan Albatross (*Diomedia immutabilis*) in 32 days and another two from Washington State, a distance on the great circle of 3,200 miles in 10 and 12 days. As far as is known, these long journeys are not regularly undertaken by birds of these species and the question naturally arises as to how they do it.

It is known that many birds including homing pigeons improve their homing performance with increasing acquaintance of the terrain; but such learning would only be relevant to the last stages of the above journeys. The fact that many long-distance migrants breed in successive years on the same nest site is at the later stages more akin to a process of contextual or perceptual learning than one of navigation and it is probably in this category that examples of *Ortstreue* belong. It is conceivable that prevailing winds with different characteristics of temperature and humidity could be helpful and so, too, the swell lines in the ocean; but many journeys by birds have crossed different climatic zones. On the long ocean flights, the position of the sun is changing throughout the day and, if used as a reference, would require a coordinated system of use.

A variety of theories have from time to time been proposed to account for the navigational abilities of birds. The influence of the earth's magnetic field as a factor must be balanced against the results from several experiments in which birds were placed in strong magnetic fields before being released, or subjected to the attachment of powerful magnets, with no significant effect upon homing ability. Again, the suggestion that some reference to the Coriolis force due to the earth's rotation might be relevant—that a body moving north or south of the equator would encounter a smaller sideways component —seems hardly applicable. If such a force were detectable, the speed and weight of the bird would have to be maintained constant within limits so fine as to be unattainable. Further, the influence of Coriolis force on the volume of liquid in the semi-circular canals of birds would be too small to have any appreciable effect. From an evolutionary point of view, the evidence is also negative. Thorpe and Wilkinson (1946) have shown that the semi-circular canals of regular homers and migratory birds are, as a class, not significantly larger than those of other birds.

Vision is a very important sense for birds. The immense speed of light, which permits rapid detection of information while the bird is moving at speed, is essential for coordination both in predation and avoiding it. There is much evidence that birds have very good vision and are in many ways better equipped to detect movement than are human beings. As Pumphrey (1961) has pointed out, cones (cells mediating colour vision) in the bird retina are less concentrated in the foveal or central area of the retina than is the case with human beings. Secondly, the greater part of the surface of the bird retina lies in the image plane so that objects at various distances can all be in focus. Much more of a varied landscape is thus clearly visible to the bird immediately. Again, birds do not have the matrix of blood vessels which serve to nourish the inner layers of the mammalian retina. Instead, they have a roughly conical pecten with folds on its surface which covers the point where the optic nerve enters the eye. The apex of this cone points forward towards the pupil of the eye. Menner (1938) indicated that the folds of the pecten throw shadows on the retina, thereby tending by the fluctuation of shadows associated with a moving object to maximise the animal's attention to movement, a feature which could also be enhanced in Pumphrey's (1948) view by the steep contours of the fovea of the bird retina. Pumphrey's evidence was that the degree of folding of the pecten tended to be related to the bird's mode of life. The heaviest folding and presumably the most extensive range of shadows was typical of hawks. Birds feeding by day on insects came next, and nocturnal birds had the simplest pattern of all. Again, Maturana and Frenk (1963, 1965) have demonstrated in the pigeon retina the existence of ganglion cells with specialised functions. One type responds to the edge of small objects moving in one direction but not in the reverse. A second type responds to the movement of an edge moving up or down in a vertical plane. Such cells may mediate the approach of many bird species to the rotating sectors described by Smith (1962) and Smith and Nott (1970).

But whatever the explanation, the direct evidence from the studies of Leibowitz (1955) and Boyce (1965) has shown directly that the avian eye is much better at detecting movement than the human eye and can in fact detect movement as slow as 15° per hour from a point of reference.

Evidence of navigation by birds

From actual studies of migrations there can be no doubt that many species of birds can navigate successfully without the assistance of older birds. Young Cuckoos appear to have managed quite well for many generations, and experiments reported by Rowan (1946) with Prairie Crows (*Corvus brachyrhynchos*) in Canada and Schüz (1949) with White Storks (*Ciconia ciconia*), which delayed the departure of young birds from breeding grounds until after all other birds had left, indicated that the great majority of the young birds had travelled in the normal direction, albeit with a wider variance than in the customary migration. White Storks taken from nests in the Baltic area, reared and released near Essen in West Germany, were in the main recovered to the south-east which was the normal direction for their population and distinguishable from the normal direction (south-west) of the local population of storks. The direction taken appears to be innate and little influenced by topographical and climatic factors. A later study by Schüz (1950) in which 754 young White Storks from the Baltic were released in West Germany while the migration of local storks was in progress, however, demonstrated the considerable influence of older birds because the recoveries indicated a predominant trend to the south-west. Nevertheless, there are several studies in which young birds have been reared and released in areas where their species does not breed. These include those of Schüz (1938) who reared 21 White Storks from the Baltic in England, Bloesch (1956, 1960) who reared 192 White Storks from Algeria in Switzerland, and Vaught (1964) who reared 377 Blue-winged Teal (*Anas discors*) ducklings from Minnesota in Missouri. Recoveries indicated a trend in the direction normally taken by the population from which the young birds derived. The experiment by Perdeck (1958) is an example of another approach. Migrating Starlings (*Sturnus vulgaris*) were trapped and displaced laterally from Holland to Switzerland while they were actually in the process of the autumn migration. Recoveries indicated a clear trend for the young birds to maintain an approximately west, south-west direction, parallel to the normal direction. It was noted, however, that older birds, as a class, tended to be nearer the area (southern England and northern France) to which they would normally migrate, possibly as a result of previous experience. More of the younger birds adhered to the customary direction and several were recovered in the Iberian peninsula. It has also been noted in several displacement experiments that the displaced birds not only maintain a direction roughly parallel to the normal route but are also recovered from approximately the same distance from the point of release, possibly as a result of genetically determined, hormonal factors which would control the duration and strength of the migrating tendency. While local features of topography and general suitability for feeding and nest sites would probably determine the terminal areas of many migrations, a definite innate component seems likely.

One interesting and puzzling feature observed of many released birds is that termed by Matthews (1961, 1963, 1968) 'nonsense orientation'. Mallards released from Slimbridge on the Severn estuary in south-west England, for example, within a few seconds of taking off, orientate in a generally north–west direction. This direction is not maintained for more than 20 minutes but is maintained irrespective of the point of release, time, sex, age, experience or wind direction. Mallards taken from London parks also showed their own peculiar orientation tendency, to the south-east. The phenomenon is found in several other species of birds. Griffin and Goldsmith (1955) and Goldsmith and Griffin (1956) found that Common Terns (*Sterna hirundo*) from nests on the Great Lakes and the New England coast of North America flew south-east on release which, as in the above examples, would confer no apparent, biological advantage. A possibly comparable 'nonsense orientation' has also been reported of the flightless Adélie Penguins (*Pygoscelis adeliae*) of the Antarctic by Penny and Emlen (1967). Those from the Cape Crozier rookery moved north, north-east even when transported hundreds of miles. In the Antarctic, with wind and pack-ice moving generally from east to west, a generally northward movement from any point on the mainland would have some biological advantage; but it is of interest to note that the Adélie, although apparently disorientated when the sun is completely obscured, are capable of maintaining a compass direction when released under clear skies and, even though they set off in a north, north-east direction, were capable of returning considerable distances southward to their nests. Some remarkable journeys were the return of 18 out of 20 birds a distance of 192 km in a time period of 8 to 30 days, and of 20 released on the Ross Ice Shelf, 9 were back on their nests within 12 to 36 days, a distance, over difficult terrain, of 300 km.

The ability of many birds to navigate would appear to have been established beyond reasonable doubt and some further discussion of the possible methods is in order. Perhaps the experiments of Kramer (1949) gave the clearest, initial insight. Kramer had noted that, at times of the year corresponding to normal migration, many passerine birds in cages showed increased restlessness and a directional trend in their movements. This led to a number of experiments by Kramer and many others in circular cages with a variety of devices for recording the birds' orientation. It was found (Kramer, 1951) that a starling placed in a circular cage was disorientated under heavily overcast skies but could orientate in cloudy conditions, provided the sun's approximate position was apparent. By placing mirrors in his circular apparatus which changed the apparent position of the sun through 90°, the starling's 'nonsense' direction was changed correspondingly. Similar results were obtained in later experiments (Kramer, 1952) with starlings trained to a particular direction and so, too, with pigeons (Kramer and Reise, 1952). Matthews (1963) has also reported an observation relating to the 'nonsense orientation' of Mallards which in the south-west of England normally fly to the north-west. The sun was setting in the south-west, but in the heavy cloud a break appeared in the north-west through which the red afterglow was reflected, giving the false impression that the sun was setting in the north-west. At this time, the Mallards released flew to the north-east.

There seems little doubt that the sun is the most important reference for the orientation of birds; but its position changes and for its effective use, the birds would need a sense of time. This they appear to have although the precise basis has still to be clarified. It probably derives from the bodily rhythms associated with the cycle of day and night. It has been possible in experiments to 're-set' these internal clocks. Hoffman (1954) for example trained two starlings to take a particular direction with reference to the sun. The birds were then kept for 12 to 18 days under artificial conditions of light and dark, which gave an artificial 'day' six hours behind the normal day. They were then tested under natural sun at times when its height was the same as would be expected in their artificial day, namely, 9 a.m. artificial time and 3 p.m. normal time. Their orientation changed clockwise through 90° from the direction in which they had been trained and under conditions of constant light was maintained for 23 days. It was reinforced by training under the retarded time system for 29 days. It was apparently found possible to 're-set' the birds' clock, this time forward to 'natural' time, by housing them in an outside aviary for 12 to 16 days, when the birds resumed their original orientation. Comparable experiments have been performed by Schmidt-Koenig (1958) who trained Pigeons to a particular direction and then changed their clocks plus and minus 6 hours and by 12 hours in summer. The new orientations were in the predicted direction. Matthews (1963) was also able to alter the 'nonsense

orientation' of Mallards in the expected, general direction by sub-
jecting them to artificial 'days' of advanced and retarded time.

There is also evidence that many species of birds can orientate at
night. Sauer and Sauer (1955), using the Kramer-type cage, showed
that Garden Warblers (*Sylvia borin*) and Blackcaps (*Sylvia atri-
capilla*), although not hitherto exposed to the night sky, would
orientate in the appropriate autumn direction for the local population. Sauer (1957) showed, too, that the same birds revealed a re-
versed direction in the spring. Many caged birds have manifested
nocturnal orientation consistent with their migratory movements
at the time of testing; but it has been noted that the orientation
is best on clear, starry nights and very poor under overcast
conditions. By fixing small lights to the legs of Mallard, Blue-
winged Teal, Pintail (*Anas acuta*) and Canada Geese (*Branta
canadensis*) Bellrose (1958, 1963) has shown that 'nonsense orientation' is the same by night as in the day. Matthews (1963) has
in fact shown that the same individual Mallards could orientate as
well under starry skies as by day; but again all of these studies with
free-flying, wildfowl at night indicated that the birds were disorientated in overcast conditions. Mallards with 'time-shifted' internal clocks, however, still flew in the customary 'nonsense' direction (north-west) as did controls when released under starry skies.
Under conditions of a full or half-moon on clear nights, 'nonsense
orientation' was poorer than on clear moonless nights, a result which
many workers have reported of caged migrants. On the other hand,
Matthews (1968) has reported that in conditions where the stars are
obscured by overcast but the position of the moon is apparent, the
customary north-west orientation of Mallards can be observed, and
with the moon in different positions. Such conditions are difficult to
predict and experimental evidence will thus be slow in forthcoming,
but much more may be learnt of the reference to the moon in
nocturnal orientation.

All the evidence indicates that many species of birds make use not
only of celestial reference points but also of topographical features
and local landmarks. Indeed the best of human navigators, fresh
from a trans-ocean flight, often has to rely on local enquiry and information or the services of a taxi-driver to find a particular address.
The traditionally accepted homing abilities of Pigeons, while by no
means as impressive as feats reported of the Manx Shearwater or the
Laysan Albatross, owe much to what can only be learning and
memory. For successful homing flights by racing Pigeons, a number
of training flights, increasing in distance and from various directions,
are necessary, and even then losses are reported which increase
noticeably in foggy or overcast conditions.

ECHO-LOCATION

The foregoing accounts of orientation and the homing abilities of animals, impressive though they are, should not be closed without some reference to some special cases. Bats have very poor vision and are nevertheless capable of flying at considerable speeds, avoiding obstacles and capturing insect prey in conditions of poor daylight and at night. Researches initiated by D. R. Griffin in 1938, when he approached the physicist G. W. Pierce at Harvard University who at that time was studying sounds above the frequency range of human hearing, indicate that a sound echo-location system is used. With the bats (*Myotis lucifugus* and *Eptisicus fuscus*) it was possible to establish that they emitted high-frequency sounds of short duration. This opened up a rapidly developing field of research, the early stages of which have been admirably described by Griffen (1958, 1960).

As often happens in the history of science, forerunners of important advances emerge on closer study of the problem and it appears that Lazaro Spallazani (1729–99), born in Modena and later a professor at the Universities of Reggio, Modena and Pavia, had done some crucial experiments. Quite late in life, he noticed that blinded bats were able to fly quite effectively and avoid obstacles. He also noted that an object hanging in front of the mouth caused disorientation; but one of his most rigorous experiments was to blind 52 bats which he had captured from a roost in the bell tower at Pavia and return early in the morning four days later to capture the bats as they were returning from hunting. He caught and killed four of the blinded bats and found that their stomachs were just as full of insects as those of the normal animals. Griffin and co-workers were able to demonstrate that bats 'blindfolded' by painting over the eyes with black collodion showed no disorientation. Those with plugged ears or tightly sealed lips, however, were disoriented. Those with one ear covered or plugged never crashed head-on into a wall, but they were not successful in avoiding fine wires. A slight crack in the sealant applied to the lips enabled the bats to orientate and there is evidence that families of bats may vary in the method of sound emission. For the family Vespertilionidae the mouth is the normal channel of sound emission. The Long-eared Bat (*Plecotus rafinesquii*) uses mouth and nostrils and the horse-shoe bats, the Rhinolophidae, the nose. The type of frequencies emitted by the bats is related to their mode of life. The common insectivorous bats of North America and Europe emit frequencies of the order of 90 to 40 kilo cycles per second although higher frequencies, possibly harmonics, have been recorded. The more experienced and skilful bats consistently used more of the higher frequencies, while lower frequencies were more in evidence among bats emerging from hibernation, fatigued bats or young, inexperienced bats. The frequencies emitted began at a high level and fell towards the end, a decline of 90 to 40 kilo cycles being common. By positioning microphones ahead and at 90° to the head of the bat, it was established that the intensity of the pulse recorded at the side of the bat's head is lower than straight ahead.

There is thus reflected back to the bat a whole matrix of directional cues which is the basis of their remarkable orientation, most of which is done in the dark or in the poor light conditions of dusk. Different species of bats can in fact be distinguished by their patterns of flight which varies with the insects upon which they consistently prey. Griffin was able in 1951 to record the pulses of two species of bats as they were in the process of attacking insect prey over a small pond near Ithaca in New York State. The smaller of the two species *Myotis lucifugus* conveniently came later, hunting by skimming the surface. *Eptisicus fuscus* came while there was still light and flew straight towards the parabolic horn, detecting apparatus. This rather narrowed the field of reception but was sensitive. By 'translating' the frequencies downwards, the high frequency emissions were made audible and it was evident that as the bat closed on the prey or the pebbles, buck shot or coarse sand thrown in its path by the experimenters, the rate of the clicks emitted accelerated until they were as short as one millisecond. The cruising pulse for these species is 10 to 15 milliseconds in duration and of relatively low frequency, ranging from 34 kilo cycles at the beginning to 32 at the end, although *Eptisicus* are capable of emitting frequencies beginning at 70 kc and ending at 30 kc. The use of cruising pulses is clearly a useful adaptation since rapid emissions while cruising could give rise to confusion in the reflections from a variety of objects. In later studies in the laboratory with bats of the genus *Myotis* (*Myotis lucifugus*, *M. subnulatus leibii* and *M. keenii septentrionalis*), Griffin et al. (1960) divided the hunting activities into a search phase, an approach phase and a terminal phase, each marked by a progressive shortening of the inter-pulse interval, and a noticeable drop in the frequency of the pulses emitted in the terminal phase. There is evidence that the sounds of insects themselves may attract the hungry bat in some conditions, but the system of echo-location seems to be predominant. Audible, short clicking sounds or 'ticklaute' are used by *Rousettus aegyptiacus* which are vegetarians and have vision. These sounds increase in intensity as the light is diminished.

A similar method is reported of the Oil Bird of Caripe (*Steatornis caripensis*). In 1953 Griffin visited the great cavern occupied by the Oil Birds (*Guacheros*) at Caripe in Venezuela and found the interior so dark that a film exposed for 9 minutes did not differ from an un-exposed film. A sound-recording apparatus set up in the entrance to the cave recorded a steady stream of very sharp clicks of short dura-tion (1 to 1·5 ms) and with a frequency range of 6,100 to 8,750 cycles per second which is well within the range of frequency audible to human beings. An experiment was conducted with four birds which had been captured in a trammel net. It was noted that the clicks stopped when a flashlight was pointed at the birds. In a room 10 × 12 × 8 ft high, the birds flew well enough in conditions of light and darkness although it was noted that the wing tips frequently brushed the cord of the electric light. The ears of the three strongest birds were plugged with cotton wool and duco cement. Each bird flew directly into the wall but orientated successfully, once the plugs were removed. With the lights turned on, the birds flew competently; but it was noted again that they emitted fewer clicks. The Oil Birds were not as good as the bats in the rapid detection of fine objects, as would be expected since the bats use high frequencies from 30 to 100 kilo cycles, involving short wavelengths of ·3 to 1·1 cm, while the lower frequencies of the Oil Birds would involve wavelengths of the order of 5 cm.

Even more remarkable than the bats' use of echo-location is the method of evasion of bats evolved by the noctuid moths with a very simple sound-registering device in the form of a pair of ultrasonic 'ears' found near the 'waist' of the moth between thorax and ab-domen. This very simple 'ear' of the moth nevertheless appears to be useful. Roeder and Treat (1960), by observing 402 field 'incidents' between moths and bats, arrived at the result that for every 100 moths which reacted to the bats' presence and survived, there were 60 survivors among the moths which did not react.

Returning to the concept of echo-location, the experiments of Schevill and Lawrence (1953) and Kellog (1958) show that the shallow water porpoise or Bottle-nosed Dolphin (*Tursiops truncatus*) uses a sound-ranging and reflecting system to locate underwater objects. Reactions to vibrations of the order of 80 kilo cycles per second were indicated. In Kellog's experiments in an outdoor pool, in water so turgid as to prevent vision, an object as small as a single BB shot dropped into the water would elicit an underwater beaming response. The dropping of half a teaspoon of water onto the surface in quiet conditions would elicit exploratory sound pulses which were not repeated since no echoes were returned. Small, solid objects lowered quietly into the water elicited no immediate pulses, but were usually encountered in the normal, ranging activity. However, if the object was dropped with a splash, an immediate pattern of pulses was observed and if echoes were returned, the beaming continued.

Further experiments showed that a porpoise was capable of learning very fine discriminations in the positioning of preferred types of fish in choice experiments and avoiding a solid but invisible plexiglass door in a steel net.

A later study by Pilleri and Knuckey (1969) suggests that some Delphinidae may orientate with reference to the sun. They noticed in the western Mediterranean that dolphins rarely came to ride or play in the bow wave of the expedition's yacht unless it was travelling along the course of the sun, i.e. east–west or west–east. When on a northerly or southerly course, very few dolphins approached the yacht and if they did, they stayed only a short time. The famous case of 'Pelorus Jack' (Alpers, 1960), a female *Grampus griseus*, was apparently an exception. Over 24 years, she regularly met the ferry steamer from Wellington (North Island, New Zealand) and accompanied it back to Nelson (South Island) through the islands of Marlborough Sounds, adding one more item to many accounts of cooperation between dolphins and man in which Lamb's (1954) account of the help afforded to fishermen by the Amazonian Dolphin (*Inia geoffrensis*) must be included. Dolphins have several times come inshore off the coast of New Zealand, and played for hours with children at flicking or bunting a surf-ball. Pilleri and Knuckey noted that *Delphinus delphis* swim in the morning towards the rising sun from west to east. In the afternoons they swim towards the setting sun, whereas little sustained, directional swimming was observed in the middle of the day. It was during this directional swimming that formations were observed (see Fig. 7). The proposed investigations of the navigation of *Delphinus delphis* at night, in fog and on sunless days will be awaited with interest.

It will be apparent from the foregoing brief account that many animal species have remarkable powers of finding their way about. While the homing abilities of fish cannot be regarded as entirely proven, in view of the high rate of loss, and more rigorous statistical evidence is awaited, the remarkable feats of homing and navigation of bees, birds and bats offer evidence of independence of action, variability and retention and thus evidence in favour of purposive behaviour.

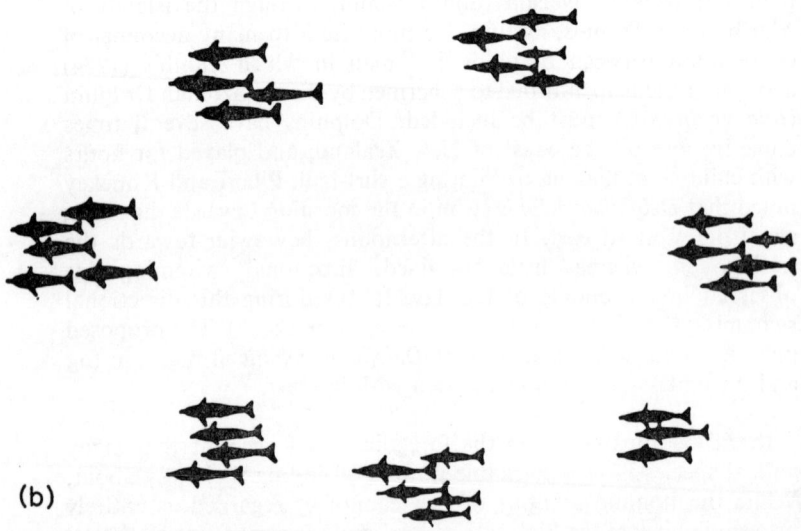

Fig. 7

Different formations observed of Dolphins—schematic representations.
(a) *Delphinus delphis* and *Stenella styx*. (b) 'Hollow circle formations of
Tursiops truncatus. (c) 'Parade' by *D. delphis*. (After Pilleri, G., and
Knuckey, J., *Zeits.f. Tierpsychol.*, 1969, 26)

5

SOCIAL ORGANISATION

THE SCATTERED COMMUNITY

The social organisation achieved by any species in a particular habitat is intimately bound up with the activities of other species, communication within and between species, density of population and considerations of territory. In the view of Wynne Edwards (1962), social organisation has a basically important function in providing competition within the species and thus enhancing optimum dispersion and survival. A selective process is clearly at work in the competition for central breeding areas among the Black Grouse (*Lyrurus tetrix*) (Kruijt and Hogan, 1967), the Ruffs (*Philomachus pugnax*) (Hogan-Warbug, 1966), Red Grouse (*Lagopus lagopus scoticus*) (Jenkins et al., 1963) and the Kittiwakes (*Rissa tridactyla*) (Coulson, 1968). Those males which do not succeed in setting up a breeding area tend not to reproduce and, as a class, are found to suffer a higher rate of mortality. So, too, a dominance order among herd animals, which tends to restrict reproduction to the senior and dominant males, is selective and probably ensures that the available females are fertilised by selected males. In Wynne Edwards' view, it does something more, namely, protect the food resources from being exploited beyond the point of recovery by obliging those ousted in the competition to disperse farther afield and substituting another form of competition, in the case of breeding sites for example, competition for space instead of direct competition for food. It is a stimulating hypothesis supported by considerable erudition (Wynne Edwards, 1962) and has implications for human societies. The crucial issues are whether or not the thesis can be shown to apply with comparable force to the social organisation of all species and if alternative explanations of the adjustment of numbers to the food supply, such as limitation of clutch size in birds to environmental indications of the food supply (Lack, 1954) and alternative forms of predation or differential mortality in poor seasons, would not be relevant. Crook (1965) emphasises as factors in the dispersion of birds the distribution of preferred foods within the environment and selection of nest sites in relation to predation.

It should be recognised that social organisations differ in complexity and rigidity and may vary in the one species from time to time. Crook (1966) has reported that in poor feeding conditions, the large (up to 400) herds of the Gelada Baboon (_Theropithecus gelada_) in Ethiopia tend to split up into much smaller 'one male' groups and 'all male' groups which come together again when food supplies are abundant. Schaller (1963), too, has noted, as have several workers with other primates, that mature males could leave the group of gorillas for a day or two at a time and rejoin. It is probable, however, that individual members were not without reference in the form of trails, dung, scent, sound and signs of recent browsing, to the position of the group, just as lone wolves, bears and domestic dogs have reference to the position and condition of others of their species and can modify their behaviour accordingly. Whatever the result, few animals in natural conditions can behave for long without some reference to others of their species.

In an interesting paper, 'The communal organisation of solitary animals', Leyhausen (1965) has drawn attention to the possible social significance of interactions between individuals who may be widely separated in space. His data is drawn from observations of domestic cats in rural Wales and the cities of Bonn, Zürich, Hamburg and Paris (Leyhausen and Wolff, 1959), and use is made of Hediger's (1949) concept of first-order homes and second-order homes. Domestic cats normally do not control their own numbers, their first-order home or the food supply; but nevertheless a picture of a community emerges from the study. Individual cats have a 'territory', a first-order home, such as a room or special place in a room, and a 'home-range' which consists of a number of places for resting, sunbathing or keeping watch which are regularly visited and connected by a system of pathways. Beyond this is an elaborate system of paths leading to places of congregation, of courting and of hunting. There is an order of dominance. A superior cat may visit the first-order home of an inferior without violence; but usually such territories are respected. When two cats meet at the conjunction of two pathways, the dominant cat may take precedence; but there is evidence that conflicts are avoided by one cat waiting or backing away or according priority to the one arriving first. At nightfall, there is a quaint meeting which Leyhausen could only describe as a 'social gathering'. Males and females come to a meeting place not far from their home-range where they sit quite close, in some cases close enough for mutual licking and grooming. There is only occasionally evidence of hostility such as the flattening of an ear or hissing. Leyhausen was able to make close observations on the Paris population and it is clear that the meetings occurred outside of the mating season and usually broke up quietly when the cats retired to their sleeping quarters around midnight.

It was noted, however, that on occasion the meeting lasted all night, possibly an indication of the approach of the mating season. The interesting feature was that these quiet and friendly meetings were attended by cats who on other occasions were observed to be fighting or chasing one another. Young males were sometimes incited to come out of their range and fight, and in this way appeared to gain admission to the 'brotherhood'. On the other hand, there was some evidence of cooperation. One cat might use a particular floor for running and playing in the morning, another in the afternoon. In uncrowded conditions, the community is not normally characterised by absolute ranking with one dominant male preventing others from reproduction. Instances were noted of females who remained faithful to a particular male of inferior rank, both in free-ranging and cage conditions from one heat to another, for several years. Leyhausen describes such a system as the relative type of ranking which appears to be associated with the checks and balances deriving from several males of approximately equal strength such that no one individual could afford to attempt to subdue the others in turn. The absolute ranking arises only when one individual is so strong as to preclude challenge. It was also noted in caged populations that there was a direct relationship between density of population and the type of ranking; the greater the density, the more evident was the absolute ranking and the less apparent was the permissive or relative type of ranking. Very crowded conditions were characterised by the emergence of a despot, greater hostility and greater neuroticism.

The concept of a community clearly permits of a range of interpretation. In one sense, practitioners of a specialised scientific discipline, scattered throughout the world, form a community. They may only meet occasionally and correspond rarely; but much of what they do in professional life has some reference to what others are doing, have done and might say, and at international conferences there is evidence of a rank order with amusing instances of submissive behaviour.

Again, many of the large raptorial birds, Eagles, Condors and

Hawks, by their hours of soaring flight, must provide a distinct, visual reference for others of their species as do the male Fiddler Crabs by waving their huge claw. Distant reference by audible communication is provided by the song of many birds, the roaring of rutting Deer, Alligators, Lions and the screaming of Howler Monkeys. Perhaps the most remarkable examples of auditory communication are those of sub-marine species. Water is a better medium for the transmission of sound than air. Sound travels 4·5 times faster in water than in air and often in the ocean there are strata or horizontal layers of water of approximately the same temperature which transmit the sound without dissipation in the third dimension. As a result, sounds can be transmitted in the ocean for considerable distances. The detonation of a few pounds of explosive near Dakar on the western coast of Africa, for example, can be recorded in the waters off Bermuda. Accordingly, many marine species have a very effective means of communicating over long distances. Perhaps the most arresting example is that of the Humpback Whale (*Megaptera novaeangliae*). From a vertical array of hydrophones in waters off Bermuda, Payne (1969) was able during the northward migration in spring to record the 'songs' of the whales which can be rendered at considerable depths, from two to four thousand feet. The remarkable feature is that the 'songs', unlike the songs of birds, last for several minutes and are then repeated, very much in the same way. Because of the excellent conduction of these sounds in the horizontal layers of the ocean, it would thus be possible for a whale to communicate with other whales over considerable distances, probably up to several hundred miles, and if the 'songs' have any characteristics typical of a particular phase of the reproductive cycle, the animal's position and breeding condition could be advertised. A lone whale would thus not be a solitary animal but a member of a scattered community.

Olfactory methods of referring to others of the species are legion. Burchardt (1958) from a study of Red Deer (*Cervus elaphus*) in the Swiss National Park near Zernez draws attention to several methods of communication over long distances including auditory methods and bodily posture. Olfactory cues, however, can be an effective means of reference when visual and auditory methods are out of range, and the preorbital scent glands ('glandes de rut') near the eye, from which small branches are marked, and the interdigital gland in the feet provide a ready means of indicating the recency of the animal's presence, its direction, and possibly also its identity. A similar practice with the pre-orbital glands has been observed in Thomson's Gazelle (*Gazella thomsoni*) by Walther (1964b) (Fig. 8). Wynne Edwards (1962) has also drawn attention to the practice of

Fig. 8

Marking by the Thomson's Gazelle from the pre-orbital gland. (After the original by Walther, F. R., *Zeits. f. Tierpsychol.*, 1964 21)

Cervus elaphus, the Red Deer of Scotland, of using particular rubbing places, such as fallen trees or a small tree which in time is reduced to a stump. These are frequently sniffed by the stags during the rut and may be used as points for urination. The scent glands adjoining the anus of the male Badger provide a ready form of reference to the home range of this sometimes lone animal. Again, the practice of several mammals, such as the Spotted Hyena, several species of antelopes, the Indian Rhinoceros and the African White Rhinoceros, of dunging in particular places could fulfil something of the same purpose, of indicating to individuals and groups the order of proximity of members of its own species. The individual thus tends to act with some reference to other members of its own species which may be distributed over a wide area and in some respects all of these animals constitute a community. Another feature is the behaviour with reference to members of other species. In terms of influence upon behaviour, the effect from other species may be more decisive than from the same species and interesting 'sociograms' reflecting the differential influence of relevant species could in some environments be worked out.

MORE COMPACT COMMUNITIES

The methods of reference to and communication with others of a species must obviously change when members are in close proximity or confront each other, when the behaviour of others is of more immediate relevance to the individual and when problems of encroachment, concession and the assertion of rights are intensified. Some remarkable examples of apparently peaceful spacing of individuals have been noted. Gudger (1949) has provided illustrations from Rotorua, New Zealand, and the Brule river, Minnesota, of trout which rest in regular horizontal rows, like soldiers on parade. From time to time, individuals withdraw for a foray at the minnows but they return. The practice of ranking may be related to conservation of energy, the arrangement with head facing upstream facilitating breathing and possibly reducing the force of the stream on individuals. Trout commonly wait below small rills or rapids and in this way have some of their food brought to them. The distance between individuals may be related to the distances which individuals of the species will normally tolerate. Portman (1961, p. 65) notes that with many deep-sea fish the distances within the shoal are relatively constant. He also cites an observation by Pitman (1931) of Crocodiles below the Murchison Falls in the Nile:

If you look down into the water on a calm day, you will be amazed at the regular placing shown by row on row of crocodiles: like warships they lie at intervals of about fifty feet and with three hundred feet distance between rows, which stretch from one bank to the other.

Again it may be largely a matter of conservation of energy in relation to fluid dynamics, as may also be the case with the large flocks of oyster-catchers and other sea birds standing regularly spaced and facing into the wind; but it is possible that the distance between individuals which members of a species can comfortably tolerate is relevant. It is also conceivable that aspects of visual communication enter into the various phases of station keeping when animals such as birds, fish or dolphins are in motion. Pilleri and Knuckey (1969) report swimming 'on parade' by dolphins in the Mediterranean. *Delphinus delphis*, for example, in calm conditions and on the west–east–west course of the sun's arc maintained formation at slow speeds of 4 to 5 knots (Fig. 7c). Bio-acoustic records of such parades are reported to be negative. In a species which has been observed to ride the bow-wave of a destroyer moving at 32 knots and leap forward from it, such parades are remarkable as is the positioning of several dolphins (*Delphinus delphis*) in raising a wounded dolphin to the surface where it blew two or three times then dived (Fig. 9).

McBride (1967, 1969) has also demonstrated with populations of feral, domestic fowls that there is a marked consistency at particular times of the year in the space permitted between individuals and that there is a further constraint of behaviour which is related to the

Fig. 9

A helpless dolphin *D. delphis*, wounded by an electric harpoon, is raised to the surface where it can breathe. (After Pilleri, G, and Knuckey, J., *Zeits. f. Tierpsychol.*, 1969, 29)

order of dominance. Outside of the breeding season, for example, a dominant male inhibited all aggressive behaviour among females within a radius of 8 to 10 ft and among subordinate males within 15 to 20 ft. McBride (1968) distinguishes between the space which the animals displace physically and an area of 'social space' around them. This 'social space' he conceives in three main categories, the fixed territory, the 'personal sphere' extending around solitary animals (or small groups) in home-ranges, and the 'personal field' of gregarious animals. The personal sphere is an area extending all round an animal while the personal field he regards as an area around the face and extending further in front than to the sides. The existence of these personal areas (both sphere and field) can only be inferred from observation when other animals infringe upon them or exceed the limits of personal distance. McBride cites an experiment in which a flock of seven male turkeys were housed in a pen which was progressively reduced in area to approximately one square metre. The birds then stood around the fence, facing outwards.

Their personal field was thus kept intact even though the personal sphere had suffered intrusion.

Clearly much detailed work could be done in this field and possible variations between species and within species for different seasons determined. McBride has evidence, for example, that intrusions into the social space, involving repeated alerts and agonistic (associated with fighting) behaviour can induce tension and stress, particularly in crowded animals. There is evidence, too, from Palmer's study (1941) that overcrowding in nesting colonies can involve intrusions into the social space of the birds with the result that an increasing amount of time and energy is taken up by agonistic activities with an associated, high incidence of reproductive failure. This field of enquiry has implications for problems of overcrowding in human communities where cultural factors would probably be important variables.

The concept of social space is basic to all forms of social organisation in animal communities and is associated with the three important inter-related problems of any society, namely, dominance or precedence, social activities associated with reproduction and conceptions of territory. For purposes of exposition, they can only be treated separately.

DOMINANCE AND PRECEDENCE

In a personal communication, Mr Roger Ewbank describes a frequent observation when a group of domestic pigs, hitherto strangers to one another, are first assembled in a pen. For about twenty to thirty minutes there is often pandemonium and even carnage and, with the abatement, it is evident that a rough form of hierarchy or dominance order has evolved in the assertion of which more aggression may be evident; but the attainment of a dominance order is associated with a lowering of the total amount of aggression in the community. Dominance hierarchies have been observed in many animal species including the familiar peck order of domestic hens, first reported by Schjelderup-Ebbe in 1922, the various species of Grouse, Seals and Sea Elephants at breeding time, Wild Cattle, Wild Horses, many species of Antelope during the breeding season, and several of the Primates.

However, it would be a very wasteful development if the hierarchy had repeatedly to be established by incessant fighting. Challenges are inevitable; but it is a more economic arrangement to assert the dominance, once established, or any priority by some form of communication which will be effective if threat is matched with some form of appeasement. The accompanying illustrations from Schenkel (1948) (Figs. 10, 11) indicate some patterns of communication in the

(a)   (b)

Fig. 10

The wolves' anal 'face'. (a) normal or neutral position, (b) expression of
superior wolf in social interactions. (After Schenkel, R., *Behaviour* 1947,
vol. 1)

Wolf. As might be expected in a species where individuals are quite
capable of killing weaker members, wolves have very effective appeasement gestures. A fierce fight, as with large domestic dogs, can
be brought to an end by one contestant crouching and exposing its
throat or fleeing with hindquarters lowered and tail between the
legs. They are not pursued very far. Another device is the 'belly-up'
submissive gesture (Fig. 12). The animal lies on its back or side but
teeth and claws are not presented in the attacking position. If they
were, the animal, while forfeiting movement, could still do harm
since all its weapons are in a position to be deployed against the
aggressor; but in the submissive gesture, they are withdrawn and the
head is turned away. Schenkel has suggested that such a submissive
posture may derive from infantile attitudes such as rolling before the
mother to be nuzzled, playful sparring with the mother while lying
on the back, and crouching associated with suckling and begging for
food. It is of interest that Walther (1965) in noting the lying down
position as a submissive gesture in various species of gazelle, in Kudu
and Wildebeest, also suggests that the pattern derives from attitudes
common in infancy when the young, in response to danger, lie down

D

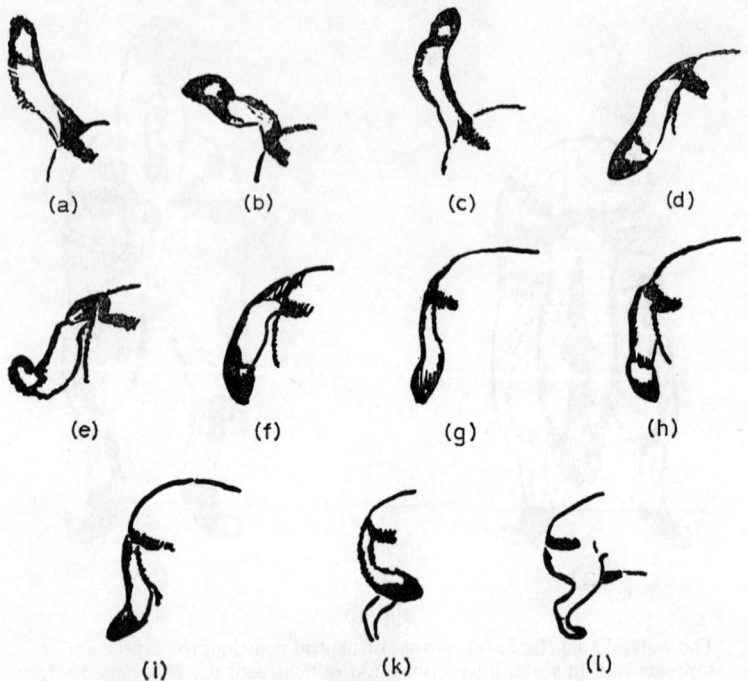

Fig. 11

Various tail positions adopted by wolves—indicating: (a) confidence in social intercourse; (b) positive threat; (c) with sideways movement—intimidation; (d) normal position in situations without social tension; (e) partial threat; (f) similar to (d)—particularly frequent while observing and eating; (g) depressed mood; (h) between threat and defence; (i) with sideways wagging, active submission; (k) and (l) strong restraint and submission. (After Schenkel, R., *Behaviour*, 1947, vol. 1)

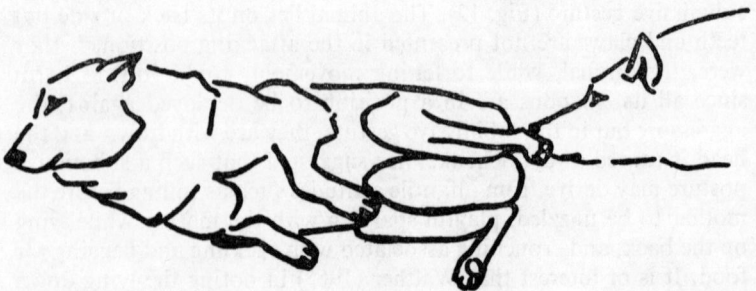

Fig. 12

Passive submission by one wolf to another. This position is often adopted by young males and females. (After Schenkel, R., *Behaviour*, 1947, vol. 1)

in grass or thicket as close to the ground as possible. In the Dorcas Gazelle (*Gazella dorcas*), the appeasement gesture resembles the lowered head position of a calf approaching the mother to suckle (Fig. 13) while the opposing threat posture consists in the head and

Fig. 13

A dominant, male Dorcas Gazelle with high horn presentation, confronts a less dominant male who adopts a submissive pose. (After the original by Walther, F. R., *Kosmos*, 1966, vol. 62)

horns held erect, and high above the submissive animal. There are probably many gradations of threat postures which can reflect the animal's mood. The high head position facing the intruder may indicate vigilance and readiness to resist and a lowering of the horns, immanent attack. Turning the horns to one side, on the other hand, would suggest a pause in hostilities.

Geist (1966) concluded from his fine study of the Bighorn Sheep (*Ovis canadensis*) that the chief factor in determining the male dominance hierarchy was the size of the horns. Where the difference in size was considerable, conflicts were reduced; but fierce, head-on butting contests between rams with horns of comparable size occurred, the challenger being afforded the advantage of charging downhill. By analogy with the 'peck-order' of domestic hens, the more dominant ram kicked those beneath him in the rank order with one of his forefeet and this was accepted without reply. Mounting of the subordinate was also observed. In a later paper, Geist (1968) has pointed out that several other characteristics of the Bighorn Sheep differ with age. The small, horned sheep tend to be lighter in colour than the large sheep, both on the back and on the belly and rear side of the legs. The almost pure white underbelly and rear leg margins of the young sheep becomes brown in the older sheep. The

progression of size and horn development is indicated in Fig. 14 and
it appears that horn and body size are more influential in determining social ranking than sex. The large, sexually mature males interact
at will with all sheep but associate most with males of comparable
horn size and the larger females, whereas younger males and juveniles
interact in the main with sheep of equal or smaller horn and body
size. Geist suggests the apparent emphasis upon size may be in part

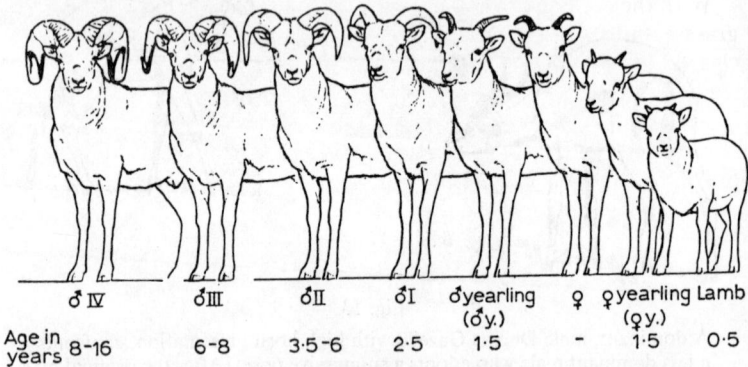

				♂yearling	♀	♀yearling	Lamb
♂ IV	♂ III	♂ II	♂ I	(♂y.)		(♀y.)	
Age in years 8-16	6-8	3·5-6	2·5	1·5	-	1·5	0·5

Fig. 14

The sex and age classes of the Bighorn Sheep showing the steady increase
in horn and body size from the lamb to the class IV ram. Note that the
adult ewe and yearling ram are almost identical in appearance. ♂=male,
♀=female. (After Geist, V., *Zeits. f. Tierpsychol* , 1968, 25)

due to the lack of obvious sexual characteristics in this species. The
kick with the forefeet on subordinates is, however, likened to the
'Laufeinschlag' noted by Walther (see p. 121), and the sniffing of the
vulva and 'Flehmen' posture (wrinkling of nostrils after sniffing
vulva or immersion in urine) is common in many ungulates.

The importance of horns in ranking orders has been further emphasised by Espmark's (1964) study of semi-domestic and free-
ranging Reindeer (*Rangifer tarandus*) in Swedish Lapland. During
the rutting season the mature bulls were at the top of the ranking list
and, in a separate experiment, the cutting of the antlers was associated with a distinct drop in the hierarchy; yet the ranking was not a
straightforward reflection of antler size and condition. There was
evidence as in studies by Allee (1952) of triangular formations (A
dominates B, B dominates C; but C dominates A) and of fluctuations
with seasonal changes in the animal's physiological state and of learning. Allee, for example, found that mice which had been castrated
definitely fell in the hierarchy; but if they had had experience of fighting before castration, they very often continued to be aggressive.

In Espmark's study of reindeer, the top rank was held by a bullock in which, as a castrated animal, the usual hormonal factors would not apply; but the animal had once been used in domestic haulage jobs and the influence of familiarity with man's ways and greater confidence could be considered. Size and strength were also found to be important between sex groups. During the rutting season, mature cows were very aggressive to young bulls but became more tolerant with the onset of winter.

With the coming of the snows, the reindeer change their food from grasses and sedge to lichens, which are exposed in a series of craters cleared by the animals digging with their forefeet. Usually the bullocks are the first to initiate the digging, and around the craters a new dominance order emerges which again may be associated with age, size, the possession of antlers and physiological factors. The older bulls, which are in poor physical condition after the rut, drop their antlers in early winter. The barren cows and yearlings drop them in late winter and pregnant cows with an attendant calf retain them until after calving. In the new order, the pregnant cows are among the highest ranked members on the grazing areas and in view of the nutritional needs of the foetus and the calf, which follows its mother until the new calf is born, the biological advantages are obvious.

Frequent changes in rank order were also noted by Schloeth (1961) in a study of the cattle living in a semi-wild state in the Camargue in southern France. In a particular herd of 26 animals upon which attention was concentrated, there were, in the period from 15 May 1955 to the end of August 1958, 6 changes of rank among the 10 bulls, 13 among the 12 cows and 2 among the 4 calves. It was noted that during the rutting period there was much chasing and very little bodily contact. In winter, the position was reversed. Low-ranking animals may solicit bodily contacts, such as licking, touching horns, snout and forehead and sham fighting, from those of higher rank; but it was noted that there was a preference for any exchange of bodily contacts to be confined to the next three ranks, up or down, and 45% of such activities in fact took place between animals of adjoining rank. Territorial behaviour and issues of precedence were observed in grazing areas, drinking places and in favourable places for rest and rumination. Certain expressive movements, such as pawing the ground, rubbing the head and neck on the ground and thrusting the horns at the ground were used as threat postures when dominance was challenged, followed by fighting and chasing, if the challenge continued. Flight is in effect appeasement, since the vanquished are rarely pursued very far. No such expressive movements were ever addressed to a member more than three ranks distant in the hierarchy and, among bulls, 67% of such expressive movements were addressed to an animal with whom a change of rank subsequently occurred. The figure for cows was 60%. Castrated males were observed to chase cows more frequently than the bulls and were not necessarily lowest in the male rank order. As in many animal societies, the young were observed to form separate play groups and at times initiated play activities among the whole herd.

A remarkable ritual which has the effect of providing opportunities for appeasement and reducing fighting has been reported by Walther (1965) of the Grant's Gazelle (*Gazella granti*) observed in Ngorongoro crater, Tanzania (see Fig. 15). The male rivals march towards each other, head erect. At a distance of 10 metres, the head is averted and the rivals continue, taking up a parallel position, facing in opposite directions about a metre apart. With the forequarters of one opposite the rump of the other, they raise their heads high, and together turn the nose inward, displaying the conspicuous white patch on the throat. This may be repeated several times. In 47% of the cases observed, one of the rivals gave up the display by turning the head away from his rival and walking away. In 11%, fighting followed from the display and, in 12%, one of the bucks jumped on the other from behind and fighting followed. The net effect is nevertheless a reduction in fighting; but the display does provide a ritualised method of asserting dominance and also of accepting it. Yet another ritual has been reported of the Zebra (*Equus quagga*) by Backhaus (1960). The intimidation display (see Fig. 16a) and threat (Fig. 16b) can be followed by some particularly fierce fighting, involving biting and that powerful kick from the hind feet with both forefeet braced on the ground also used by zebra when pursued by predators. The fiercest of encounters can frequently be brought to an end, however, by the submissive gesture of one male placing his head over the rump of the other (Fig. 16c), a posture also used by females in soliciting attention. Pfeffer's extensive study (1967) of the Corsican Mouflon (*Ovis ammon*) also provides clear evidence of postures of dominance and submission and a peculiar gesture of appeasement in which the dominant animal kneels and is licked on the neck by the less dominant male. Very probably, the existence of the many forms of ritualised combat does lessen the total amount of damage. During the ritual struggle, one party can form an impression of the strength of the other and break off the struggle with less damage than might be incurred by horned animals in indiscriminate fighting. The ritual horn-pressing of the Lesser Kudu bulls (*Tragelaphus imberbis*) (Fig. 17) and the ritual 'neck struggle' of the male and female Greater Kudu (*Tragelaphus strepsiceros*) (Walther, 1964a) (Fig. 18) are but two of many examples.

Among the Chacma and Hamadryas Baboons (Hall and De Vore,

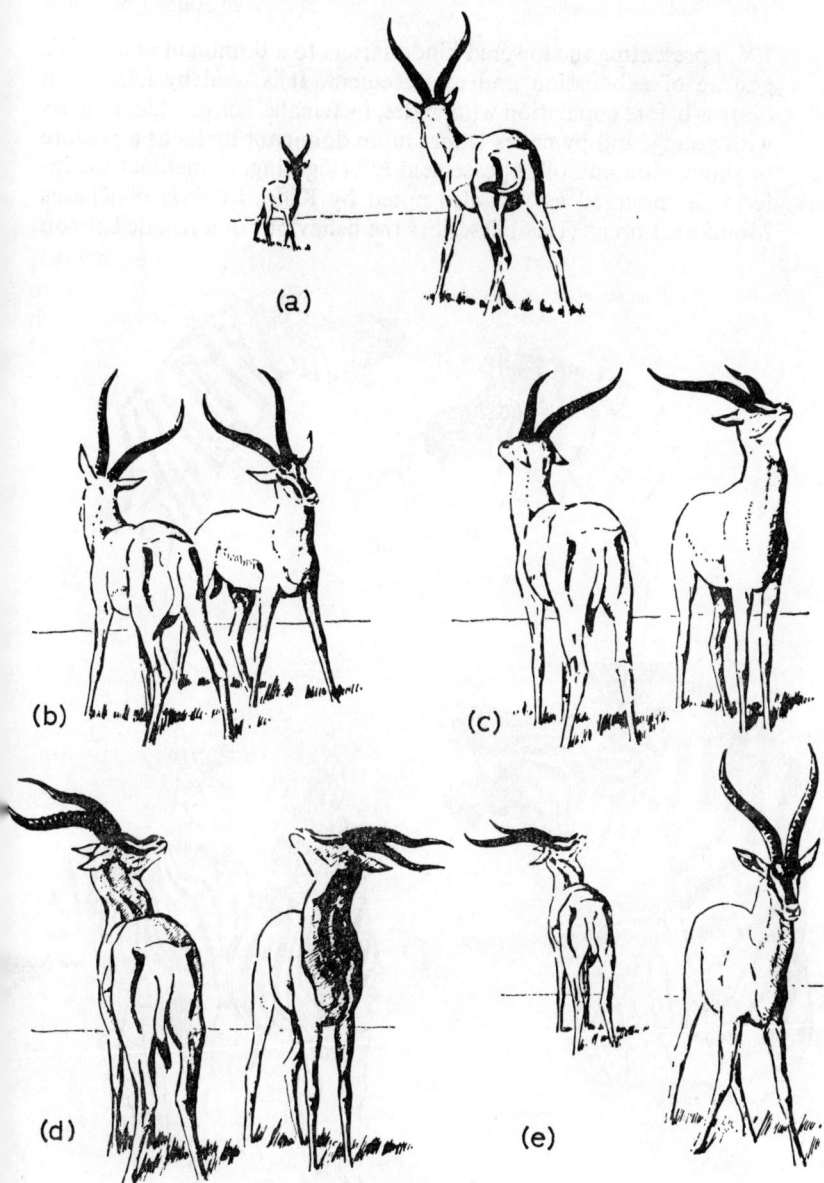

Fig. 15

Phases in the display procedure of the Grant's Gazelle. Phase (d), which exposes the white throat patch, may be repeated several times. In phase (e) the 'winner', or more dominant male, continues the display. (After the original by Walther, F. R., *Zeits. f. Tierpsychol.*, 1965, 22)

1965) presenting the lowered hindquarters to a dominant animal is a gesture of submission and appeasement. It is used by females in oestrus before copulation with males, by females before older females with young, and by males before more dominant males as a posture of submission and of appeasement after fighting. Sometimes the inferior is mounted as was also noted by Koford (1962) of Rhesus Monkeys. Lorenz (1966) describes the behaviour of a female baboon

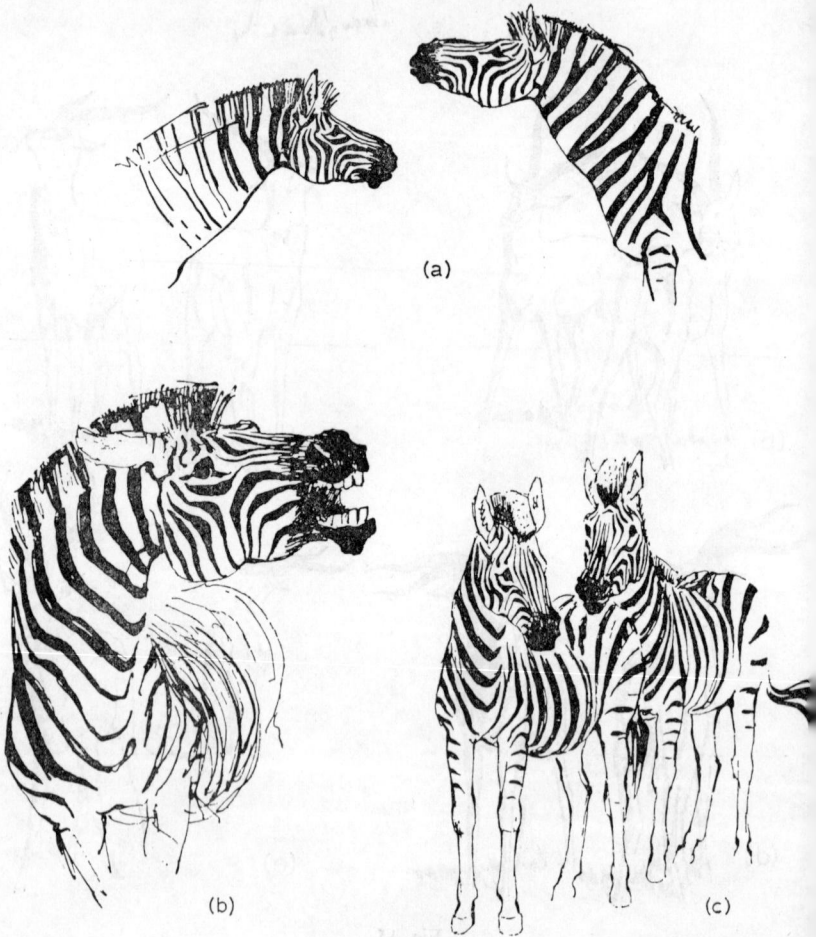

Fig. 16

(a) Intimidation display of the zebra (*Equus quagga*), (b) threat, (c) submissive gesture after fighting—similar to affectionate approach of the female. (After Backhaus, D., *Zeits. f. Tierpsychol.*, 1960, 17)

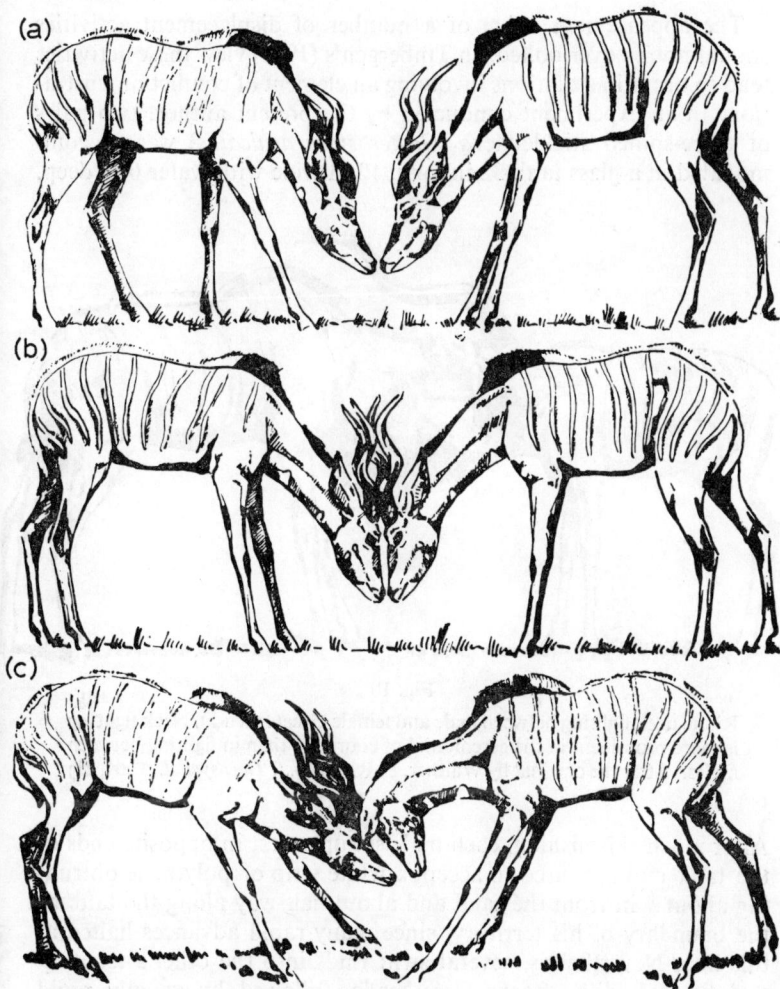

Fig. 17

Ritualised struggle between two Lesser Kudu bulls. (a) nasal 'examination', (b) rolling and pressing of foreheads, (c) horn pressing. Other fighting techniques are also used by the Lesser Kudu bulls, such as locking the horns and frontal pushing. (After the original by Walther, F. R., *Zeits. f. Tierpsychol.*, 1964, 21)

which had been largely reared among human beings, on first being admitted to a strange room. The animal presented her lowered hindquarters before every chair in the room, probably because in the strange context they were frightening objects.

The appeasement value of a number of displacement activities should not be overlooked. In Tinbergen's (1951) view these activities tend to occur in situations involving an element of conflict or frustration. In an experiment conducted by the present author, two pairs of three-spined sticklebacks (*Gasterosteus aculeatus*) were accommodated in a glass tank 32 in long, 12 in wide with water 6 in deep.

Fig. 18

Ritual neck fighting between male and female Greater Kudu. Neck fighting is more prominent in the ceremonial of courtship than in fighting between males. (After the original by Walther, F. R., *Zeits. f. Tierpsychol.*, 1964, 21)

After much skirmishing, each male set up a nest in opposite ends of the tank and appeared to accept a white strip of polythene obtruding about $\frac{3}{4}$ in from the sand and about half-way along the tank as the boundary of his territory, since many rapid advances halted at this line. Nevertheless, several swift raids into the other's territory and fierce bodily attacks were made, followed by equally rapid retreat. Two contending motivations appeared to be at work, aggression against a rival male and fear or timidity at venturing out of one's own territory, to which conceivably could be added the fear and inhibition associated with attack from the rival. As had been noted by Tinbergen, displacement digging with snout burrowing in the sand occurred on both sides of the boundary and each day several craters appeared, the distribution of which could be altered by moving the polythene boundary a few inches either way. The net effect was probably a reduction in the number of aggressive attacks which could have some biological value since, at the end of four or five days of such attacks, both males were swimming around with strips of flesh trailing from their sides.

In an analogous way, fighting European starlings may pause and preen their feathers. Tinbergen has also noted that Herring Gulls engaged in fierce combat may suddenly stop and begin to pluck grass as they do when nest-building. Again, domestic roosters may suddenly suspend hostilities and peck at the ground. All of these displaced activities occur in other contexts and it may be that in a highly excited animal, when the normal manifestation of energy is inhibited, an overflow of energy into other neural circuits occurs, as Makkink (1936) implied with the term 'sparking-over' movements or activities. The probable net effect is, however, a reduction in the damage associated with aggression.

The foregoing brief account will have shown that many types of community are possible and that the form of the community, as in the Gelada Baboon or the colonial developments by sea birds in the breeding season, can change over time. There is evidence, too, that such factors as overcrowding can influence the degree of despotism and neuroticism. Many communities including scattered communities do indeed provide a means for ensuring conventional competition with obvious advantages for survival of the species. The consistency of the social organisation, including the ranking orders, communication and rituals of threat and appeasement in each species from one generation to another would not occur without some element of selection and genetic transmission of behaviour patterns. It will be apparent, however, that none of these consistent patterns of behaviour could operate without some awareness by the individual of the behaviour of others and instant response to any subtle variations. No hierarchical system and pattern of communications could function without some form of retention or 'memory' by the individual and in the establishment and maintenance of such organisations as well as in all aspects of maintenance, such as predation and its avoidance, food seeking and travel, the individual must be capable of independent action.

6

RITUAL IN SEXUAL APPROACHES

There is no lack of evidence from anthropological and sociological studies of human societies of patterns, conventions and rituals which in varying degrees and at times strictly guide the approach of the sexes. The field of comparative behavioural and ethological studies could almost certainly provide a wider range of examples and many of them of greater exactitude. Indeed a complete account would attain encyclopaedic dimensions with ample evidence of stimuli or signs which release successive stages and data indicative of reciprocal communication.

The courtship sequence of the three-spined stickleback from the initial approach of the male to the gravid body of the female or models thereof, the construction of a nest by the male, the beautiful zig-zag dance by male and female in parallel, the driving or inducement of the female into the nest, the butting of the spine of the female to assist in the emission of the eggs, the subsequent fanning or aeration of the eggs by the male and his subsequent care of the young have been so often observed and experimentally studied by Tinbergen and collaborators as to leave no reasonable doubt that the pattern which unfolds derives from innate sources. Many other examples could be cited of comparable consistency, adding weight to the ethologist's case that the classification of animal species on the basis of behaviour patterns would be at least as good as the conventional method using bodily characteristics.

While only a few examples can be mentioned here, the behaviour of the male fiddler crabs of the many species of *Uca* as recorded in the studies of Pearse (1913), Verwey (1930) and Crane (1941) attract attention. In the main, the fiddler crabs inhabit the area uncovered by the tides where both males and females make individual burrows from which they emerge at low tide to feed on the small organisms in the sand. At this time, the males indulge in their well-known practice of signalling with their giant claw which is of little use in feeding and appears most relevant to signalling and also fighting, to which all fiddlers are prone. Females fight among themselves and males sometimes fight with females; but the fiercest encounters occur between the males. Both males and females appear very much attached to their holes and fiercely resist encroachment by others. The remarkable feature is the variation in the shape, colouring and manner of waving the huge chelipeds. In *Uca latimanus* there is an enrichment of colour in both claw and carapace on emerging from the burrow. In waving, which serves to attract females and threaten males, the distal end of the claw is turned toward the body and the varied movements are reminiscent of a violinist's bowing arm. Crane (1941) from a study of the behaviour of twenty-seven species of *Uca* at Panama recorded (pp. 152–3) that

each species proved to have a definite, individual display, differing so markedly from that of every other species observed, that closely related species could be recognised at a distance merely by the form of display. Furthermore, related species had fundamental similarities of display in common, and series of species, showing specialisation of structure, in general showed similar progression in display.

When the female has been attracted, there may be a period of familiarisation; but copulation once begun is resumed at high intensity. It is conceivable that the elaborate signalling may assist in bringing the female (and the male) into a state of readiness at about the same time.

A similar story showing differentiation into distinguishable poses and at the same time clear affinities with postures of other surface feeding ducks has been reported of Mallards by Lorenz (1958) (see Fig. 19). Numbers 1 and 10 occur independently in the Mallard whereas the combinations 4, 3 and 5, 6 occur in all the surface-feeding ducks. Crosses between different species have produced new combinations of motor patterns, recognisable from the traits of the parents, and suppression of others, as well as movements recognisable in other species but not in the parents. One conclusion is that the response was latent in one of the parents. The work provides clear evidence of the genetic basis of such specific movements and opens the door to the close study of the evolution of behaviour. It is important to remember that such behavioural traits would not have evolved unless they served some biologically useful purpose and that whereas it is often possible for the human male and female to avoid or short circuit many rituals of courtship, and reproduce, this is probably more difficult among many species of animals.

The detailed work of Tinbergen and collaborators over many years has provided a wealth of information on the ritual postures of the gulls among which the Herring Gull (*Larus argentatus*) is of special interest because of the variation in the method of pair formation. Many Herring Gulls are already mated before they arrive on the breeding grounds and these are the older birds, one at least of whom has mated in a previous season. There appears to be appreciable mortality, but otherwise the pairing lasts for life although the evidence of Tinbergen (1953) suggests that the pairs are not together during the winter but apparently exercise their remarkable powers of visual and auditory recognition to associate again in the spring. Instances are cited of a bird sleeping on its territory amid the myriad cries of the flock and awakening instantly at the cry of the returning mate. Younger birds, which are readily distinguishable by the retention of some of the juvenile colouring in their spotted secondaries and tail feathers, return to the breeding grounds and associate in various 'clubs' of unpaired birds. Here the initiative is taken by the females which approach the male with neck outstretched, bill pointed forward and an occasional toss of the head. They utter subdued calls and eventually circle the male several times. The male may react in one of three ways. He may make himself as tall and broad as possible and adopt the threat posture towards other males. If they are receiving attention from females, fights may occur. In the second reaction, the male stretches the neck forward and utters the 'mew' call of the species. He then walks a few yards with the female to a place where he makes incipient, nest-building movements in which the female joins. In a third and less common way, the male stretches and twists his neck in various directions to which the female responds by increased head tossing and walking to and fro in front of the male who appears to try to avoid her attentions but does not retreat. After a time, he regurgitates with difficulty a mass of half-digested food which the female begins to peck and swallow greedily even as it is emerging from his beak. Tinbergen's impression is that as the pair bond

Fig. 19

Courtship poses found among surface feeding ducks—as exemplified in the Mallard. (1) Initial bill shake, (2) head-flick, (3) tail shake, (4) grunt-whistle, (5) head-up—tail-up, (6) turn to female, (7) nod-swimming, (8) turning the back of the head, (9) bridling, (10) down-up. (After Lorenz, K. Z., The Evolution of Behaviour, *Scientific American*, December 1958)

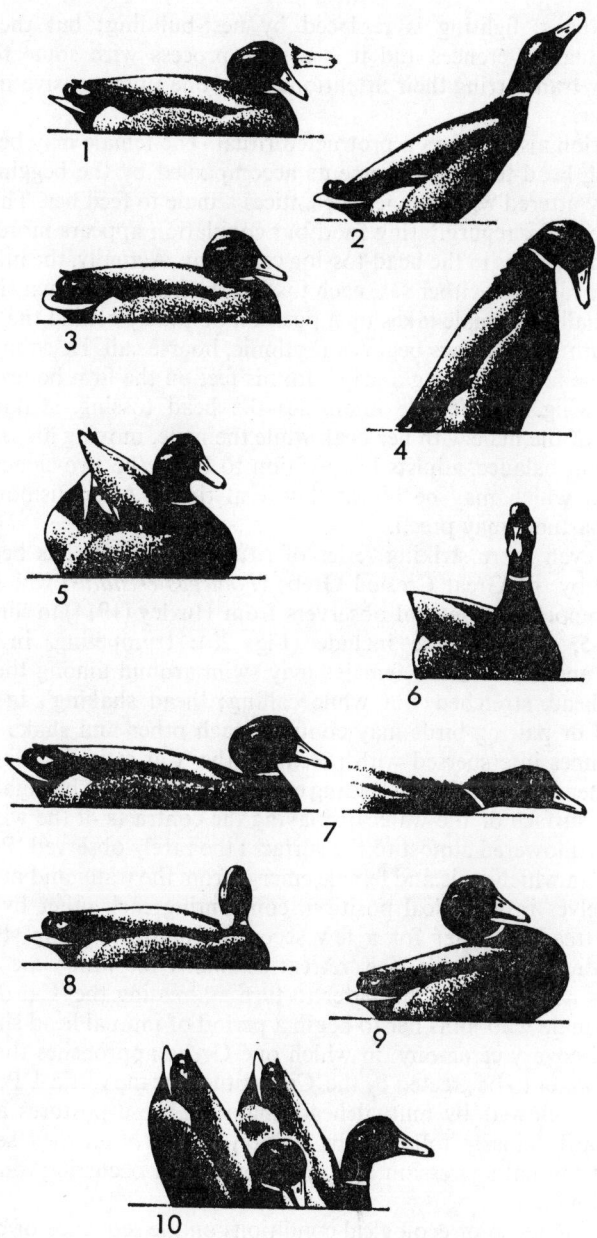

Fig. 19

strengthens, fighting is replaced by nest-building; but there are individual differences and it is a slow process with some females rapidly transferring their attentions from one unresponsive male to another.

Coition also follows a protracted ritual. The female may begin by making head-tossing movements accompanied by the begging call, usually uttered when the female entices a male to feed her. The male may react by regurgitating food but copulation appears more likely if the male joins in the head-tossing ceremony. Actually, the initiative may be taken by either sex, each tossing the head in its own rhythm. Eventually, the male takes up a position obliquely behind the female and with raised wings begins a rhythmic, hoarse call. He soon jumps onto the female, taking a grip with his feet on the firm bones of the inner wing. The female intensifies the head tossing, striking the breast of the male with her beak while the male, moving his wings to maintain balance, adjusts his position to bring the two cloacas into contact which may be repeated several times. After dismounting, both partners may preen.

An even more striking series of ritual movements has been observed by the Great Crested Grebe (*Podiceps cristatus*) which have been reported by several observers from Huxley (1914) to Simmons (1954–5, 1959). These include (Fig. 20): trumpeting, in which males and on occasion females may swim around among the reeds with head stretched out while calling; 'head shaking', in which paired or pairing birds may confront each other and shake heads, sometimes interspersed with preening; the 'Cat attitude', displayed by either sex in which the leading edge of the wings are brought down to the surface of the water displaying the contrasts of the wing and the head lowered almost to the surface; the rarely observed 'Penguin dance' in which male and female emerge from the water and maintain themselves in a vertical position, confronting each other by vigorously treading water for a few seconds while presenting strips of vegetation in the bills; the 'retreat ceremony' in which the female patters noisily across the water to turn addressing the 'Cat display' to the male who joins her to begin a period of mutual head shaking; the 'discovery ceremony' in which one Grebe approaches the other under water to be greeted by the 'Cat attitude' or the vertical 'Penguin' stance, followed by mutual head shaking. These postures are not necessarily closely followed by coition. Quite often they serve to reduce mutual aggression and fear, copulation occurring somewhat later, on land.

The influence of ecological conditions on the sequence of behaviour patterns is well illustrated in Simmons' (1970) study of the Brown Booby (*Sula leucogaster*) at Ascension Island. There is clear evidence from guano deposits and sub-fossil bones that the birds once nested on the island; but human settlement associated with the appearance of cats and rats has now confined the colonies of this large bird (wing span about 3 ft) to small colonies on rocky stacks off the coast. The diet of the Brown Booby is mainly fish captured inshore and this supply varies widely from year to year so that in some years, shortage of food prevents reproduction. The species has adapted to these conditions by maintaining the pair bond and ownership of the nesting territory throughout the year. Further, unlike the short, annual breeding season of many colonial birds, the Brown Boobies at Ascension remain in breeding condition for roughly eight months. The birds are thus able to reproduce over long periods, when the opportunity offers. Normally only two eggs are laid and hatch about five days apart. When the first nestling is a week or ten days old, it evicts from the nest the other sibling which appears to have served a certain 'insurance' function in that, had food supplies wavered in the first few days, the presence of a reserve would avoid the recommencement of the whole breeding cycle and ritual which nevertheless can be resumed during the greater part of the year if the food supply is adequate.

The year-round occupation of the nest-site by the pair is marked by many recognisable postures (Fig. 21). On coming in to land on the site, both birds utter calls specific to the sex and hover briefly with flapping wings which displays effectively the white under-feathers of the wings. There is an element of threat or assertion of ownership in the gesture and it is addressed to the mate, intruders or near neighbours if they are on or near the nest-site. The female has a bowing display, consisting of a frontal bow while calling and pecking at nest debris, which is addressed to more distant, possible intruders while the male deals with more immediate threats of usurpation by ruffling the feathers and stabbing with the bill in their direction. Both birds defend the nest-site which is occupied throughout the year, the female being generally the more violent. When a foraging mate has landed, there is frequently a greeting ceremony enacted by bill-touching. On occasion if the returning bird does not touch bills, the other may fly out to meet it or perform the bowing ritual. A male may sometimes greet the female by pointing the bill skywards, accompanied by the so-called 'wheezle-whistle'. Within the nest, following the greeting ceremonies, there can be frequent bill touching and sparring with the bills, mutual feeding in which the male takes the initiative, sometimes playing the role of the chick in inserting the beak into the gullet of the female, and among older pairs, mutual preening. Nest-building movements—picking up and presenting fragments of nest materials which have been gathered or stolen—accompanied by frequent neck

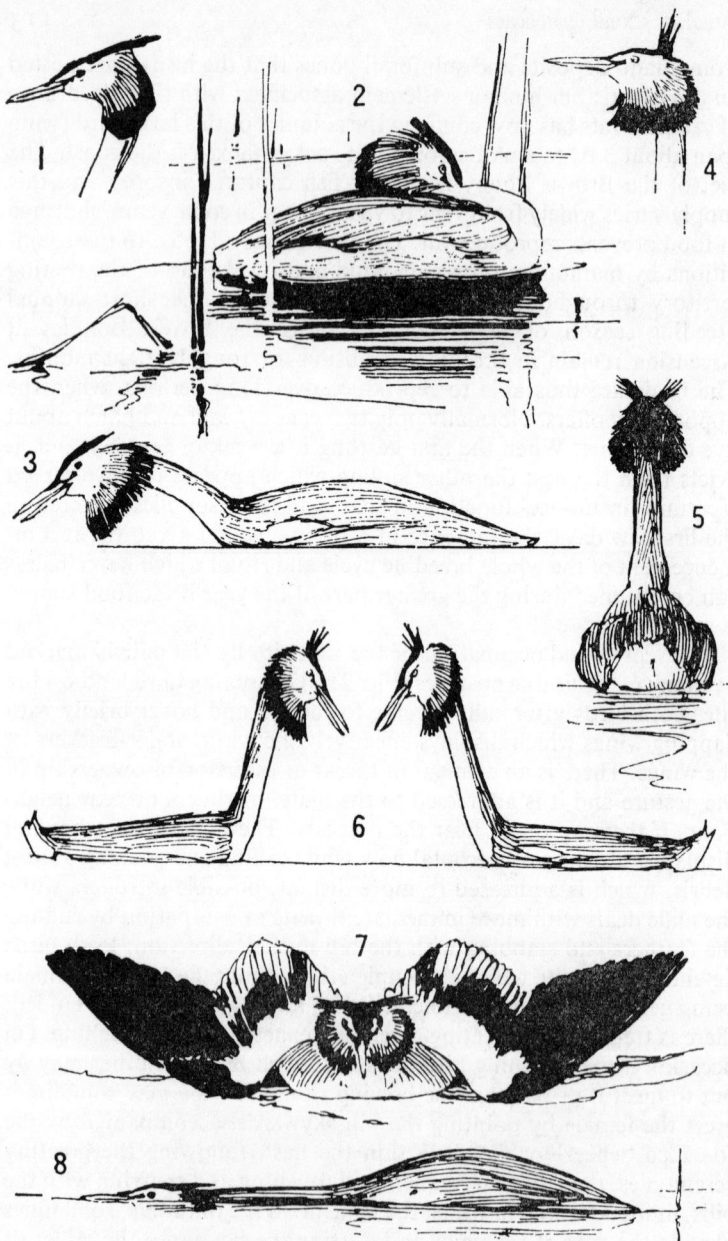

Fig. 20
Courtship of the Great Crested Grebe. The appearance of both birds is
altered not only by different bodily positions but also by movement of the
feathers, particularly those of the head and neck. (1, 2) Relaxed and
resting phases, (3) attitude of search with spreading crest, (4, 5, 6, 11)
shaking attitude from various positions, (7) the 'Cat attitude' display

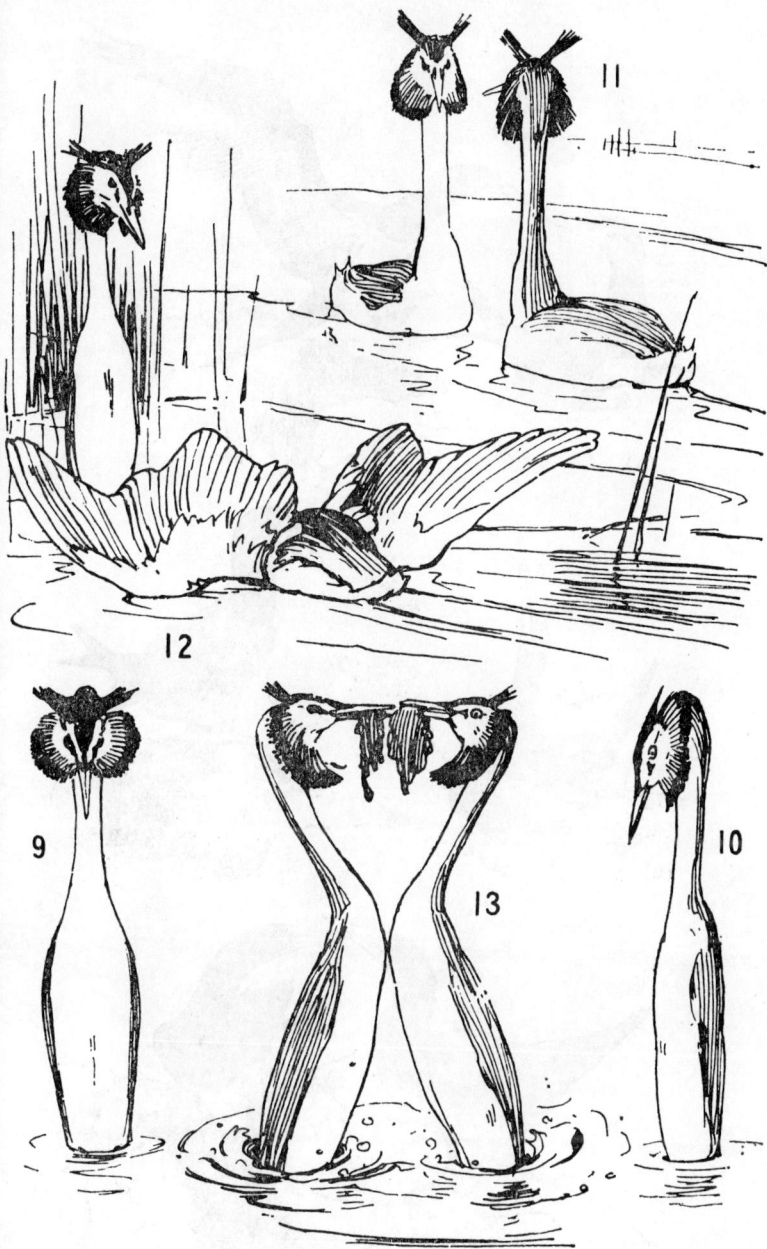

which may be adopted by both sexes, (8) the passive pairing attitude with total closure of the crest, (9, 10) the male dives and appears in the stiff upright 'ghost' position while in (12) the female reacts with the 'Cat attitude', (13) 'Penguin' dance. The pair dive and emerge, treading water for a second or two, breasts touching and presenting strips of water plants. (After Huxley, J. S., *Proc. Zool. Soc. London*, 1914, 11)

(a)

(b)

(c)

Fig. 21

Ceremonial phases in the pair-bond of the Brown Booby. (a) The greeting ceremony—in this case, the male in landing, touching bills with the female and showing the white feathers of underwings and body. (b) The greeting of the female by the 'sky-pointing salute' of the male. (c) Billing —touching and sparring with the bills—male on the right. (d) Ceremonial feeding. The male (left) plays the role of the chick. (e) Neck-crossing. The male (right) has a feather in the bill. Presenting and indicating material and neck crossing while in the nest hollow are part of the mating ceremony. (After the originals in Simmons, K. E. L., pp. 37–77, in Crook, J. H., Ed., *Social Behaviour in Birds and Mammals*, London, Academic Press, 1970)

crossing, are common. All of these movements may lead up to the mating sequence; but copulation does not occur unless the female solicits in the characteristic way, namely, by ceasing to give the usual nest calls, remaining still in the nest hollow and then 'quiver nibbling' repeatedly at a piece of nest material held on the tip of the bill. While this continues, the male mounts and copulates. The female may interrupt the mating sometimes by bill sparring, depending on the degree of receptivity.

Consistent features of ceremonial in pairing relationships are not confined to birds. Buechner and Schloeth (1965) have provided an informative study of the Uganda Kob (*Adenota kob*). This antelope has a highly organised system of territorial behaviour and the entire population of the Toro Game Reserve of Western Uganda, where the study was conducted, is orientated around some 15 territorial breeding grounds. An average ground consists of a cluster of from 30 to 40 relatively fixed territories each from 15 to 30 metres in diameter, some 10 to 15 of which are concentrated in a central area about 200 metres in diameter, where most of the mating occurs. The areas are clearly distinguishable by the worn or trampled state of the grass or ground. These central areas tend to be those which the females prefer and competition, frequently manifested in fighting, for one of these areas, which an active male may be unable to hold for more than a day or two, is particularly keen. From these central areas there is a gradient of intensity towards the periphery to territories which may be held for a year or more. Breeding occurs throughout the year and is promiscuous. Females in oestrus enter the territories where they may remain for a day, the mating activity being most intense on three or four central territories where upwards of ten females may at peak periods be concentrated in a particular territory.

The mating ceremony begins when the female enters the territory. The male usually follows behind her with head held high with short, rapid, prancing steps in a generally stiff-legged gait (Fig. 22). The tail is usually raised and the phallus erect. When the female stops, the male sniffs the vulva which stimulates a flow of urine in which the male inserts the nose. After a thorough wetting, he lifts his head and retracts his lips and wrinkles the nose in the 'Flehmen' posture. The head is moved from side to side and the lips are licked. The male next touches the underparts of the female with a stiff fore-leg— termed 'Laufschlag' by Walther (1958). This is usually gentle but varies with the state of excitement. The female then often circles lightly round the male (*Paarungskreisen*—Walther, op. cit.) and may

Fig. 22

Features from the ceremonial of mating in the Uganda Kob. (a) The approach of the male when the female enters his territory. The female is encircled with a prancing gait, the white patch on the throat being plainly visible. (b) The 'Flehmen' posture—wrinkling of the nostrils and retraction of the lips. (c) Laufschlag—more common pre-coital form. (d) Laufschlag with pincers movement—very gently executed and, more commonly, post-coital. (After Buechner, H. K., and Schloeth, R., *Zeits. f. Tierpsychol.*, 1965, 22)

Fig. 22

butt or nip at the hindquarters of the male. Intention movements of mounting usually follow. Pauses between each display are brief, if they occur, and the whole series may be completed in two to three minutes.

A further pattern of activity follows coitus. Immediately after dismounting, the male stands quietly with arched back for 2 to 5 seconds. The still erect phallus is waved jerkily and thereafter a much more variable sequence is followed including whistling loudly through the nostrils which elicits further whistles from other males across the territorial grounds. Meanwhile the female has been standing with arched back and straddled hind legs. The male may lick the vulva and the inguinal region, where small glands, exuding a waxy secretion, are situated, by pushing the head through the straddled hind legs. The udders may also be licked. The nuzzling may also be followed by further 'Laufschlag' either from between the legs or from the side in the form of a pincers movement with the male's head extending over the back of the female. This may induce circling behaviour or butting of the male by the female. These post-coital displays may be interrupted or prevented by intruders but diminish as excitement abates.

Walther (1964b) has noted many points of similarity in the territorial behaviour and mating ceremonial of Thomson's Gazelle (*Gazella thomsoni*). Urination by the female is prompted and the urine is tested by 'Flehmen'. The 'Laufschlag' gesture is common and is also found in the Dorcas Gazelle, as is the extension of the 'Laufschlag' into the pincers movement (Fig. 23). Among the Okapi (*Okapia johnstoni*) there are distinctive features and also some patterns of behaviour in the mating ceremony which have been observed among the antelopes (Walther, 1960). These include the reversed parallel position (head to tail) from which the pre-copulatory herding or circling may begin and also the 'Laufschlag'. Studies of the African Antelope of the genus *Tragelaphus* which includes the Greater and Lesser Kudu, Nyala, Sitatunga and Bushbuck by Walther (1964a) have revealed further well-developed mating ceremonies in which neck-fighting (Fig. 18) plays a prominent part. All *Tragelaphus* bulls stretch the neck in their approach to the female and in herding, they overtake from the side (Fig. 24).

It will be again apparent from the above few examples that the ceremonial behaviour patterns of related species reveal many common features with the ever-present possibility of distinctive features for those most closely related. The ethologists' case for using behaviour as a basis for classifying animal species as an alternative to a taxonomy based upon physical characteristics is certainly worthy of the most serious study. Many homologous features will be found

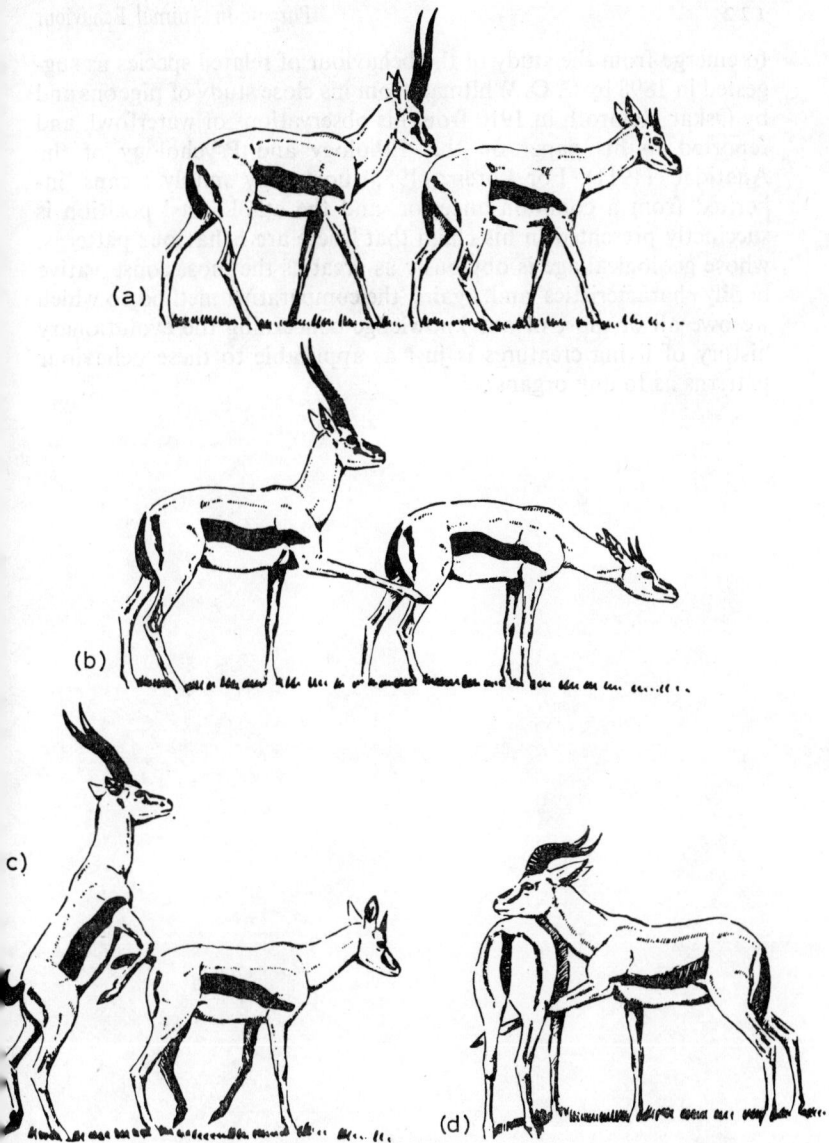

Fig. 23

Phases in the ritual pairing-march of Thomson's Gazelle. Note in (b) the Laufschlag which also occurs in a number of other gazelles e.g., Dorcas Gazelle, Mountain Gazelle, Speke's Gazelle, Persian Gazelle, Gerenuk, Dibatag, Springbuck and in diminished form in Grant's Gazelle. In (d) Laufschlag with pincers movement by the Dorcas Gazelle. (After the original by Walther, F. R., *Zeits. f. Tierpsychol.*, 1964, 21) to emerge from the study of the behaviour of related species as suggested in 1898 by C. O. Whitman from his close study of pigeons and by Oskar Heinroth in 1910 from his observations of waterfowl, and reported in his paper on the 'Ethology and Psychology of the Anatidae' (1911). For Lorenz (1955) homology simply means 'inherited from a common ancestor' and the ethologists' position is succinctly presented in his claim that 'there are behaviour patterns, whose geological age is obviously as great as the most conservative bodily characteristics' and, again, 'the comparative method to which we owe all or most of our knowledge concerning the evolutionary history of living creatures is just as applicable to these behaviour patterns as to any organs'.

Fig. 24

(a) Another feature common among the genus *Tragelaphus*—in this illustration, the Sitatanga Buck. The male, in the pairing approach, pushes the fully extended head along the back of the cow. (b) The stretching of the neck and overtaking of the female from the side during the pairing ceremonial of the Greater Kudu. (c) Mounting by the Greater Kudu bull. The head is held closely along the back of the cow. (After the originals by Walther, F. R., *Zeits. f. Tierpsychol.*, 1964, 21)

(a)

(b)

(c)

Fig. 24

7

TERRITORY IN THE SOCIAL ORDER
AND REPRODUCTION

As will be apparent from the foregoing pages, the concepts of social dominance, mating ceremonial and territory are closely interrelated and the fact that they must be dealt with separately for expository purposes introduces an artificial element.

In a much quoted paper, Burt (1943) makes the distinction between home range and territory in mammals. His view (p. 351) is that

> Home-range is the area, usually around the home site, over which the animal usually travels in search of food. Territory is the protected part of the home-range. . . . Every kind of mammal may be said to have a home-range, stationary or shifting. Only those that protect some kind of home-range by fighting or aggressive gestures, from others of their kind, during some phase of their lives, may be said to have territories.

It will be apparent that the density of population will influence the size of the home range and probably also the territory. It is known that the home range need not cover the same area during the life of the individual and that whereas the home ranges of many animals may overlap without undue difficulty, encroachment upon territories would be resisted. The awareness of individual property and particularly that of a defended territory appears to be present in many animals. It is of interest to note, as Hediger (1962) has pointed out, that the German verb to possess is 'besitzen', literally, what one can sit on, and while many mammals by scent-marking may make known the extent of their range, they resist more violently encroachment upon territory over which they have surveillance and particularly if this is used for sexual purposes.

However, it is evident that both home range and territory are part of the matrix of factors in the animal's struggle for existence and that, between the several factors, a variety of reconciliations have evolved. As shown in the previous chapter, there are clear advantages in the Brown Booby maintaining the same nest-site and being capable of reproduction over many months because of the uncertainty of the food supply. Migratory species on the other hand tend to become territorial only during the period of reproduction.

Studies of the Wildebeest by Talbot and Talbot (1963) and Estes (1969) indicate a wide range of adaptation to grazing conditions from the nomadic and migratory to the sedentary pattern of life. In the Serengeti-Mara plains, Talbot and Talbot felt that the western White-Bearded Wildebeest (*Connochaetes taurinus hecki*) was the most important single species in the plainsland complex but, with the conditions of rainfall in the area, the species tended to be no-madic and displayed many features of a nomadic species. Calves usually stand within 5 minutes of birth and within another 5 minutes are running alongside the mother. Within a day after birth the calves can run as fast as the adults. A week-old Wildebeest was actually clocked at 35 mph; all are features of considerable survival value in an area of high predation. Sometimes at the beginning of the rut, males may set up and defend territories; but the necessarily nomadic character of the pattern of life usually results in males setting up breeding herds of from 2 or 3 to 150 females and young, herded by 1 to 3 males who, when sharing the herd, appear to avoid sparring but go out to challenge and chase away other males who intrude. From time to time, old males would set up home ranges adjacent to water holes, usually orientated with reference to some central topo-graphical feature, such as an anthill upon which they may have evacuated or otherwise 'scented' from the pre-orbital glands and would challenge intruders; but these conditions only operated while the herd was stationary.

Estes (1963, 1969) studied *Connochaetes taurinus* in the rather more favourable grazing conditions of the Ngorongoro crater, Tanzania, where the nomadic pattern, typical of the semi-arid plains, existed but was interspersed by a sedentary pattern with a perma-nently established territorial network, which at the peak of the rut and in crowded conditions rivalled the 'lek' system of the Uganda Kob in density. The average distance between territorial Wildebeest in the crater was about 130 yards, but the range could vary from 30 yards to a mile according to the density of the population and the features of the terrain. The possession of a territory was virtually essential for reproduction. Away from his exclusive territory which is defended by elaborate, agonistic rituals, the male Wildebeest appears to lose many sexual and aggressive tendencies.

Various patterns of adaptation are also found among birds. With some birds, there are advantages in solitary nesting within a territory. The birds may feed near the nests; this saves much time and energy, is conducive to more intensive care of the young, and may be better adjusted to the food supply which a large colonial organisation would soon exhaust. Again, nests which are cryptic and dispersed are more difficult for predators to find. Mating is less likely to be interrupted by competition and the likelihood of the spread of diseases is reduced. On the other hand, if the food supply is adequate, there may be advantages against predation in the communal type of nesting in inaccessible places and in a departure from the usual monogamous pairing in birds, as in some species of the Ploceine weaver birds studied by Crook (1964).

The importance of the food supply has been illustrated in a recent study by Zahavi (1971) on the social behaviour of the White Wagtail (*Motacilla alba*) which are common winter visitors in Israel and during the cold rainy days of December, January and February are often hungry. In February and March, food was placed in piles about 20 metres apart along a road. Of 120 ringed birds, 100 were seen in the area more than a week after ringing, 65 actually along the road and, of these, 25 were dominant on food piles and defending boundaries around these small areas with threat displays. Five females displayed appeasement postures to territorial males, forming stable pair bonds which lasted for several weeks. When food was abundant in nearby fields, they joined flocks feeding in the fields, but returned to the territories when food was made available in them. It is noteworthy that territorial behaviour started on the first day piles were formed. On the other hand, when food was evenly distributed over an area of 12×3 metres about 100 birds were observed to be feeding calmly as a flock, as they normally did when feeding in the open fields. From this and other experiments, it seemed possible to induce territorial behaviour in these birds at will.

It is possible in fact to distinguish a great range of territorial organisations from the very loose and widely extended to the more precise and confined, and in each case an association with one or more of the major factors in survival, such as food supply, predation and mode of reproduction, which are not unrelated, can be discerned. Several writers, notably Seton (1910, vol. 11), Murie (1944), Rowan (1950), Thompson (1952) and Burkholder (1959), have reported on the extensive range of wolves and the further extension during winter months. Burkholder, working in Alaska, used aircraft and surface travel to follow a pack of wolves (*Canis lupus*) which varied in terms of sightings from 3 to 10 animals; but it was apparent that from time to time, individuals left and rejoined the pack. Over a period of six weeks, the pack moved in a clockwise direction over an area of 100 × 50 miles and covered an estimated 700 lineal miles, the greatest distance in one day being 45 miles and the daily average 15·5 miles. The average distance between kills (usually of Moose or Caribou) was 24 miles. As did Murie, Burkholder noted very little animosity between members of the pack and indeed during feeding over a kill, playful behaviour was frequently observed. Another study by De Vos (1950) on the Sibley Peninsula, Ontario, indicated circuits within an area 20 to 60 miles in diameter, according to the food available. Rowan (1950), suspecting that the number of wolves in part of Alberta and their depredations were exaggerated, asked trappers to note accurately the number of wolves or wolf tracks seen, and the direction of travel during the winter of 1948–9. The results showed that the numbers of wolves were in fact considerably below the popular impression and that the wolves followed a circuit within an area of 38 by 50 miles.

On the other hand, Thompson (1952) observed that special circumstances in Oneida County, Wisconsin, such as the presence of man-made fire-lanes, or a block of islands with good denning potential in the centre of the range, and the tendency for Whitetail Deer to concentrate in winter in coniferous swamps, were associated with a dearth of large, regular circuits; but trails through particular small sections were consistently followed when the animals were in that area. Apart from their temporary dens, wolves do not appear to set up territories which are regularly guarded or defended.

A more definite use of territory is evident in the Vicuna (*Vicugna vicugna*), a member of the camel family (Camelidae) which with other members of this family, the Llama, Alpaca and Guanaco, inhabit the high altitude country of the Andean chain. Koford (1957) was able to observe the Vicuna in an area north of Lake Titicaca in Peru and estimates that the Vicuna range as high as 16,000 feet, about two thirds of the population living above 14,000 feet, although examples of successful adaptations in zoos at much lower altitudes to sea level are known. The high winds and low atmospheric pressure favour evaporation and in the thin, dry air there are rapid falls in temperature when the sun is obscured. The thick coat of fine wool, which is much prized commercially and may lead to their extinction, is the Vicuna's defence. In the arid habitat, the grazing density is about one animal to 10 acres and the rate of reproduction is not high. In grazing over a wide range, the animals remain in discrete family bands. An average of 30 such bands gave one juvenile, 3 or 4 older females to a male. A later count gave an average of 2·9 juveniles and 5 older females; but as Koford points out, these are averages, permitting deviations. Mating takes place in the summer when the grazing is richer. Recognition of the young follows the pattern of many breeds of domestic sheep, of olfactory inspection and rejection of the calf other than the doe's own. Yearling males leave the band and tend to join male troops in which as many as 80% may be yearlings and the remainder two-year-olds and adults. From observation, these male troops may range from 20 to 75 members and tend to be larger in January and February as the yearlings are leaving the family bands. While intolerant of the sudden, close approach of others, the male troops are virtually open societies. Individuals appear to leave and join at will and without undue hostility. There appears to be no leader. The near approach of such a troop to the grazing territory of a family band is always resisted by the male leader of the family whose attacks may disperse the troop for a time. The median size of the territory is 32 acres; but in barren parts of the Puna (the high upland grazing areas) may rise to 250 acres according to the quality of the forage. Females are not readily admitted to family bands. They are very thoroughly 'nosed' by the females of the family and may be chased away with occasional spitting. The integrity of the family band is very much dependent upon the male. If he is injured and cannot defend the territory, the females disperse. When the band moves off, following the alarm trill, the male frequently follows behind and was observed on several occasions to turn and face and even approach a human intruder.

A comparable organisation is found among Grant's Gazelle (*Gazella granti*) studied by Walther (1965), in the Ngorongoro crater, Tanzania. The family or breeding group consists of the adult male, a number of females and juveniles who graze within a territory which varies according to seasonal and grazing conditions and is defended by the male. As with the Vicuna, the young unmated males form groups which have home-ranges which are not defended and in some cases overlap. Another peripheral group derives from females which withdraw from the family group to deliver their young and then join other females with young which remain in the vicinity of the male territories for a time before rejoining the male in his territory.

With the Impala (*Aepyceros melampus*), studied by Schenkel (1966b), a similar family breeding unit of a single male, a number of females and juveniles is observed but there appears to be no definite territory. The yearling males are driven off by the adult male and form all-male groups. Individual young males may eventually acquire a group of females by challenging older males or herding away such females as they can. With the Uganda Kob (*Adenota kob*) as reported by Buechner and Schloeth (1965) there is a very definite organisation of territories within the home-range, but solely for the purposes of mating. To these sites animals in breeding condition repair, but dominated in Ardrey's (1967) view by the 'territorial imperative' which is prior to the sexual motivation because it appears that little if any mating takes place outside of these territories.

A further variation of territorial behaviour is reported by Kiley-Worthington (1965) of two closely related species of Waterbuck (*Kobus defassa* and *Kobus ellipsiprimnus*) in Nairobi National Park where the two species have interbred. Since the Waterbuck rarely moves far from water, their territories on the river bank tend to have a long tenure. The length of river frontage and the density of the thickets appear to be related to the social rank and vigour of the male who usually does not achieve a territory except in less competitive, peripheral areas until $3\frac{1}{2}$ to 4 years old. The animals leave the thickets between 8 and 10 a.m. and move up to open grassland and, depending on the weather, usually return at dusk. Territories also exist on the pasture where up to 30 females and young may congregate in contact with the territorial males. Young males are not tolerated by the territorial male and form bachelor herds or may associate with a handicapped male who plays a quasi-paternal role until the young males seek territories themselves. The calves are born in the thickets by the river and may be left there when the mother resumes grazing; but usually mothers with calves in the thicket return early. The calves, as with the calves of several other ungulate species, have little or no smell in the early weeks and are difficult to find. At about $3\frac{1}{2}$ weeks, they begin to follow the mother but may have difficulty in keeping with her if she is pursued by males on the grazing grounds. At 6 months, the young have gained considerable independence and will join with other juveniles in play groups which operate about 25 to 50 yards from the herd but run to their respective mothers in a crisis. Young females as they approach puberty will leave the group of young and join their mothers. The mother-daughter relationship may continue throughout life and may result in persistent groups consisting of mother, mother's full-grown daughter, mother's half-grown young and latest young and daughter's latest young—a feature which has been noted in red deer and other ungulates.

From external observation, the boundaries of territories may on occasion be inferred from behaviour and are frequently related to topographical features. In some species there is what appears to be a definite scent-marking of territories in addition to other evidence of occupation from spoors, grazing and elimination. The male hippopotamus on coming ashore, for example, has a definite pathway or

E

set of pathways from the river bank to his grazing areas and these routes are marked by spreading dung at conspicuous points and by shaking the short tail while defaecating. The dung is thereby scattered both laterally and upwards onto bushes where, at nose level, it is unlikely to be missed by an intruder. In a study of wild Lions in Nairobi National Park, Kenya, Schenkel (1966a) noted several methods by which the lion could advertise his presence in a particular area to transient animals. Roaring by individual lions or the whole pride gives distant warning and so, too, the proud erect manner of moving and standing at conspicuous points and looking around their territory. Scent-marking is effected by sniffing at a bush, rubbing the head in it and finally squirting urine upwards into the branches. Another method is to rub the hind feet in the urine which effectively 'scent-marks' the trail.

In Schenkel's study, between 20 and 30 lions were known to live in or visit the Nairobi National Park which borders on the Masai Game Reserve where game are protected. The area within the Park was found to be divided into four different home-ranges, each occupied by a particular pride of lions. It was clear that lions of each pride did intrude upon the home-ranges of other prides but, provided they avoided contact with the resident lions, no conflict arose, although instances of mating by animals outside their range were recorded. It was noted, however, that intruders did not show the same confidence as in their own areas. A maned lion from another area was observed to return home under cover when he heard the roaring of the resident males about a mile away and, on another occasion, two intruding lionesses and a half-grown male, having killed a warthog, examined the neighbourhood for several minutes before eating. An intruding male at another time was observed to withdraw from the territory when he met the resident lion. Cubs are born in seclusion and introduced by the lioness to the pride when they are about 10 weeks old. The introduction has a marked effect upon the pride where all relationships become more harmonious and tolerance towards the cubs is general. The presence of cubs in the vicinity is associated with greater aggressiveness towards intruders and is more marked when, in numbers and relative strength, the intruder is at a disadvantage.

The foregoing is an example of an extensive home-range, any part of which may become a fiercely defended territory and it is of interest that the animals appear to behave with an awareness and respect for the territorial areas of other prides. There can be little doubt that the existence of this relatively flexible organisation does result in a dispersion of the animals over space and an overall reduction in the amount of aggression. More difficult to understand is the territorial organisation among the Red Deer (*Cervus elaphus*) in Scotland. In Fraser Darling's (1937) study, the sexes, outside of the mating season, tend to remain apart. Herds of females with juveniles of both sexes, in the main family groups, occupy a home-range which moves to lower and more sheltered ground in winter and higher in spring. The males tend to have smaller ranges on the outskirts of the female ranges but the boundaries may overlap.

With the advent of the rutting season in the autumn, adult males attempt to herd large numbers of females into territories, but the inevitable conflicts between males means that the boundaries cannot be stable. Basically, the same organisation emerges from a more recent study by Lowe (1966) of the Red Deer on the island of Rhum off the west coast of Scotland, although the segregation of the sexes, even in winter, is far from complete. In two spring counts in 1965, over half of each sex were found to be in association; but the distribution maps for March and April show a trend for the groups of hinds to be flanked by groups of stags. Where grazings are shared as a result of any intrusion by stags, the larger and older hinds tend to be dominant in their own home-ranges with yearling stags in the lower ranks. It is possible that the rigour of the climate and the need to reconcile quality of pasture and shelter are reflected in the social and territorial order and it is perhaps relevant that other ecological changes have occurred such as the removal of the sheep from the island in 1957 and an annual cull of one sixth of the population of deer, excluding yearlings. Nevertheless, a pattern of dispersion is apparent. Over the first three years of life, males tend to move further from their place of birth (av. 2·5 km) than hinds (av. 1·4 km) in the selection of their home-ranges. There is a tendency for the younger adults to graze on higher ground, particularly in the summer months when the stags graze highest of all, and for a gradual return to lower ground with increasing age.

A broadly similar social organisation has also been recorded of feral Soay Sheep by Grubb and Jewell (1966), on the island of Hirta, St Kilda. These flocks have in fact never been shepherded by man; but the ruins of stone walls and buildings remain from the former settlement which was evacuated in 1930, when a flock of feral sheep from the small island of Soay was introduced on Hirta. The stone remains have some bearing on the social organisation in that they may be used for divisions between territories, and dry stone cells, called cleits, are used for shelter. The sheep reveal a remarkable consistency for adhering to a particular home-range, usually in small flocks of about thirty, comprising ewes, lambs and young rams. Some ewes are known to have retained the same home-range for five years, and for many groups the daily pattern of movement, particularly in the summer, is remarkably regular. Rams also form home-range groups which are smaller in numbers (2–12 members) and remain together until the mating season when members tend to wander and encounter ewes on heat. The group later reforms with the same membership and on the same ground. The consistency of the home-ranges is reflected in the finding of a number of ram's skulls, but strangely no ewe's skulls, in particular cleits. Many sheep in fact die in cleits which they have regularly frequented. To date, this regular system of home-ranges has functioned without any display of aggression and appears to ensure a viable method of dispersion; but if ever the numbers should begin to tax the food resources of the island, or predators should arrive, the repercussions on the social and territorial organisation would be of interest.

THE 'LEK' PHENOMENA

A special variation of the use of territory has already been fore-shadowed in the use of specific areas by the Uganda Kob, Grant's Gazelle and the Wildebeest for mating activities. The same general features are found in the 'lek' phenomena of birds. Females in breeding condition come to specific, adjoining areas which are occupied by males, often by fierce competition with other males, and on these 'courts' the males display so that females have a choice as to which court they will enter and where they will adopt a mating orientation. The word 'lek' appears to derive from the Swedish word 'leka', to play, and can have sexual associations. The spacing of the 'courts' may vary. With the Ruff (*Philomachus pugnax*), as reported by Hogan-Warburg (1966), the courts are only about 30 cm in diameter and a metre apart. In the displays of Gould's Manakin (*Manacus vitellinus*), studied by Chapman (1935), the courts are elliptical in shape, averaging 2·5 by 1·75 ft and spaced about 30 to 40 ft apart in undergrowth. Black Grouse (*Lyrurus tetrix*) in the northern Nether-lands were found by Kruijt and Hogan (1967) to defend fixed territories varying in size from 100 to 4,000 square metres. In New Guinea, the Magnificent Birds of Paradise (*Diphyllodes magnificus*) establish courts in the jungle with each male out of sight but in auditory contact with the others. The avenues of the Bower Birds of the genera *Ptilonorhynchus*, *Chlamydera* and *Sericulus*, which are widespread in New Guinea and Australia, are in effect display courts. Parallel walls of twigs, sometimes arched overhead, may be decorated at either end with feathers, coloured stones, white bleached bones or any coloured or bright objects, some of which may be obtained by wrecking the bowers of rival males. The walls may also be decorated with a mixture of charcoal, grass seeds and saliva or the juices of fruits and berries among which the colours blue and violet are widely favoured (see Nubling, 1939, 1941). The bower is built early in the season by the male who quickly achieves spermatogenesis but the female cycle is more retarded. The male, who is also a noted songster and mimic, displays in front of the watching female before the bower, frequently holding coloured objects in his beak and the performance goes on until with the advance of the season, the insect life becomes adequate to support the young. The female then begins to assume the crouching, soliciting position and, after mating, leaves the vicinity of the bower, builds the nest and raises the brood alone.

The study of the Black Grouse by Kruijt and Hogan (1967) confirmed earlier observations that the leking territories are remarkably constant through the spring and that the same general area is used for lek displays from year to year. The males may be divided into central and marginal males, the former occupying the smaller territories near the centre which tend to be more favoured by the females and upon which the dominant cocks display more frequently and are most aggressive. There is another class of outer-territorial males which from time to time may intrude on the main leking area and may succeed in setting up a territory on the margins; but they are frequently chased away by the occupants. Females visit the lek area in April and most of the copulation takes place during the second half of April with the highest incidence around dawn. The females walk from one territory to another and are courted by the males by squatting or circling with the beautifully contrasted white tail feathers erect and frequent vocal 'roo-kooing'. Apart from an apparent preference for the central areas, it is not clear what determines the choice of the females for a particular territory. More than 85% of the copulations were effected by central males, possibly because the density of males is greater at the centre and males on central sites tend to be more experienced and responsive.

The behaviour of the Ruffs on the lek territories is broadly comparable but with some distinctive differences. The territories of each male or 'residences' are much smaller and there is a very marked sexual dimorphism. The males are larger and develop a nuptial plumage which varies from one bird to another but appears to be constant from year to year. Hogan-Warburg distinguishes two groups of males on the 'residential' area, independent males and marginal males in effect, holders of central and peripheral territories respectively. There are, too, satellite males in both areas. These do not possess a territory of their own but one or more may occupy the residential area in the presence of the occupant. Sometimes they may be attacked and occasionally courted. Satellite males frequently change hosts and visit different leks, being more tolerantly received on small than on large leks. The status of individuals within the ranks of independent and satellite males is influenced by age. The independent male begins with the status of a marginal male and with age acquires the status of a resident male and later probably reverts to the marginal status. Satellite males change status from peripheral to central as they grow older. The key to the distinction in status between independent and satellite males appears to lie in the colour of the plumage. The typical plumage of the independent male has a black or dark-coloured ruff and head tufts or a white or almost white ruff with black head tufts. The typical satellite male has a generally lighter-coloured plumage, although anomalous plumage among independent and satellite males was noted in 11% and 9% of cases respectively.

Female Ruffs normally approach the lek area by air and are usually received by the males with conspicuous displays. Their arrival may stimulate aggression on the part of the resident male either towards the satellite male, who may be driven away, or towards nearby residents. While the resident is thus absent, the satellite male usually copulates with the female. The resident male usually copulates just after he has expelled the satellite male. The choice by the female appears to be determined largely by the plumage and behaviour of the males preceding copulation. The finesse in behaviour may reflect experience and experience probably influences the females who may visit different residences and more than one leking area.

The lek phenomena clearly involve a strong selective principle which ensures that the females are in the main fertilised by the more vigorous males who are capable of gaining and holding territories and possibly, too, by the more attractive males. In studies of the Red Grouse (*Lagopus l. scoticus*) in Scotland, Jenkins et al. (1963) have shown that those males which do not succeed in setting up a breeding territory have a higher rate of mortality than the other males which is again heavily selective and an effective method of controlling density since the Red Grouse are territorial in varying degrees of intensity for the greater part of the year. It is of interest that Coulson (1968) in his studies of colonies of the Kittiwake (*Rissa tridactyla*) has shown that male birds nesting on the edge of the colony had a significantly higher mortality than those nesting in the centre. The differences for females, while mortality was again higher at the edge, were not statistically significant. Males nesting nearer the centre were, as a class, significantly heavier than those on the periphery. Clutch size, hatching and rearing success and retention of the same mate from one breeding season to the next were also found to be greater for birds nesting at the centre of the colony. Once again a selective principle is at work and as with the Uganda Kob, the Grouse and the Ruffs, it is effected through the medium of territory.

WIDER SOCIAL FUNCTIONS OF TERRITORY

A basic question arises as to why the enormous variety of complicated rituals of threat and appeasement of sexual approach and the specialised use of defended territories have evolved as they have. Since they have evolved in so many different species and in different habitats, it is defensible to speculate that they fulfil some biologically useful purpose. Evidence has already been presented which indicates that attachment of mother and young in early life is essential for survival and that in many species a period of dependence within a group is necessary to attain full emotional development and social competence. Reproductive activities and family ties clearly bring and keep groups of animals together, and predation and fear of isolation may reinforce tendencies to grouping. As indicated earlier, there probably are differences between species and even between individual members, which vary at different times, in the life space and personal space which they find optimal; but social rituals, the defence and specialised use of territory indicate another factor in motivation and this appears to be competitive or aggressive. Lack (1953) describes the intrusion of a newcomer into the long held territory of another male robin. The intruder began to sing in one corner of the territory and from a distant part of the territory, the resident male answered. The newcomer sang again and this time the owner replied from a much closer distance. This was repeated, the resident bird now replying more vigorously. On a third repetition, the resident finally released a violent song phrase from behind a bush some 15 yards from the intruder who then flew away from the territory of a bird it had not seen. Much bird song is apparently uttered not only in the competition to attract females but also to assert the rights of the resident to his territory. In many animal species, residents behave with greater assurance on their own territory, so that a period of occupation is some assurance of stability but the territory would not be retained without repeated assertion or threat of aggression.

Storr (1968), following Wynne-Edwards (1962), draws attention to the necessity of competition and aggression between members if selection is to operate and the species survive; but aggression must be controlled and this is done by a series of conventions in which aggressive threat and display are developed and destruction of members prevented by conventions of appeasement. Society is defined by Wynne-Edwards (1962) as 'an organisation capable of providing conventional competition'—and the conventional competition which is frequently a competition for territory has been substituted for the direct competition for food. The competition nevertheless does tend to distribute the members spatially and to restrict numbers to a level compatible with the food supply; but another important implication is that a function of all the conventions of a society is the control of aggression. Storr suggests indeed (p. 31) that 'society itself has evolved as a defence against aggression; and that animals and men learn to cooperate and to communicate because they would destroy each other if they did not'.

Much aggression in animal communities tends to be of the ritualised kind which abates with appeasement gestures or the flight of one of the parties. The killing of an antagonist is more common when animals live in overcrowded conditions of captivity when there is evidence of greater emotional stress and the opponent has fewer opportunities to flee.

Man is clearly a territorial animal, ever willing to assert his rights on his own property and to seek and defend his privacy. Not only is a children's ball in the front garden or a stranger in the back garden or parlour difficult to accept and frequently resented, but costly barriers are frequently erected to prevent neighbours overlooking one's property. A group of tourists in the neighbourhood is often grounds for resentment and the defence of the motherland or fatherland against the invader is one of the most effective of patriotic appeals.

Unlike other animals, Man is the only one which destroys his own species in large numbers. About 10,000,000 men were killed in World War I and associated hostilities. In World War II, apart from greater total war casualties, if the invasion of Russia is included, something of the order of six million people were exterminated in central Europe by planned methods of destruction. A few appeasement gestures, such as saluting or shaking hands with what is for most people the dominant hand, or bowing in deference or submission, are still observed; but these are only effective between individuals, not between large groups whose motivation can be influenced by mass media, and with many modern weapons aggressor and victim can be very far apart, with consequent reduction of many of the normal restraints of excess. Moreover, the greater range of human intellect and fluency of expression probably increase the prospects of opportunism whereas an animal's behaviour is more rigidly linked to and restricted by conventional gestures.

With these points in view, some consideration will now be given to the behaviour and social organisation of some of the primates which, from an evolutionary point of view, are closer to Man.

8

FEATURES OF SOCIAL ORGANISATION
AMONG THE PRIMATES

GENERAL FEATURES

The order of primates embraces some two hundred living species ranging from the tiny insect-eating Tree Shrews to Man and reveals an enormous variety of adaptations including the peculiar hand of the Aye-Aye of Madagascar with the long probing middle finger used for probing grubs from hollow branches, the large eyes of the slender Loris of Malaysia which mediate night vision, the powerful prehensile tail of the Spider Monkeys of the New World, the long arms and hands of the skilfully brachiating Gibbons of Malaysia, Indonesia and Borneo, and the enormous physique of the gentle and retiring giants of the tropical forests which in the full-grown male Gorilla may at the age of ten years attain a height of six feet and a weight of 400 lb. All of the primates including the Monkeys, Apes and Man evolved from some common ancestor. At certain stages, particular groups diverged to pursue their evolutionary fate and some have become extinct. Man did not derive from the Apes via some hypothetical missing 'link'.

Despite elements of controversy, it is perhaps useful to distinguish thirteen families arranged in four major groups with the general titles of the Prosimians, the New World Monkeys, the Old World Monkeys and the Anthropoids (Apes and Man). The Prosimians include the Lemurs, Bush Babies, Lorises, Pottos, Tree Shrews, Tarsiers, Indri, Sifakas and Aye-Aye. The New World Monkeys of Central and South America include the Uakaris, Howler Monkeys, Squirrel and Capuchin Monkeys, Spider and Woolly Monkeys, Marmosets and

Tamarins. The Old World Monkeys of Asia and Africa include the Langurs of India and Ceylon, Colobus Monkeys, the Guenons, Baboons, Macaques and Mangabeys. A characteristic feature of this group are the ischial callosities or seating pads on the rump which enable them to rest comfortably among the branches of trees. The Anthropoids include the Gibbons, the Orang Outangs, Chimpanzees, Gorillas and Man. All of these species lack tails.

All primates including Man and the new-born human infant are capable of grasping with the hand, but except in very rare instances adult Man has lost full prehensility of the feet, which with the hands confers upon the tree-dwellers the power to move rapidly along and from the extremities of one branch to another and thus a very useful degree of immunity from predators. On the finer branches, vibration giving warning of approach is readily detectable and, moreover, much of the better foliage is found towards the end of the branches. Gorillas, Baboons and Chimpanzees spend a considerable proportion of their time on the ground but Gorillas, Orangs and Chimpanzees build nests in favourable positions in trees. Many Old World Monkeys with ischial callosities will in fact sleep on their side if a safe and comfortable place can be found. Consistent with the arboreal ancestry, multiple births among the primates are uncommon. A mother can move rapidly among the branches with one infant clasped to her underside or clinging with its effective grip to her back; but more than one would be a distinct handicap. Among all primates, a dominance hierarchy is usually apparent and tends to become more pronounced in confined conditions. It was noted of a group of 100 Hamadryas Baboons on an island in the London Zoo in the 1920s (Eimerl and De Vore, 1966) that the males were so aggressive that, within a few years, more than half of them were killed. The conditions were in fact very unnatural. Normally, the male Hamadryad herds his family consisting of the juveniles and several females and guards them jealously. The group on the island contained more males than females in a restricted space, so that fighting was inevitable. A comparable situation arose in artificial circumstances on the Japanese island of Kyushu (Eimerl and De Vore, 1966), where a group of Japanese Macaques were under study. They walked from their sleeping sites on the mountain sides each morning, always in the same order with the juveniles gambolling in front and at the sides and the dominant males walking in the centre with the females and infants. At the feeding station, they always fed in order of rank and in the special circumstances of the study, feeding was preceded by a status parade in which animals of inferior rank were mounted by those of higher rank. Normally, presentation of the hindquarters is a gesture of submission and mounting an assertion of dominance.

Grooming is common in many primates and fulfils a useful biological purpose in removing ecto-parasites. Lions for example are often found to be infested with ticks on the back of the neck where they cannot scratch. Grooming does not strictly follow the hierarchical order but the more dominant animal tends to be groomed for a greater proportion of time.

The group is very important in the early life of the primates. As already indicated, it is within the group that the young animal becomes aware of status relationships and sexual practices. There are also opportunities for learning the types and localities of edible foods and dangerous situations. Several authors have referred to the shooting of two Baboons from a car in the Nairobi Game Park, Kenya. The group had long been used to cars and it was quite easy to approach in a car and observe them. Eight months after the incident, it was still impossible to approach the group by car. Although the majority of the group had not seen the incident, they were able to apprehend fear in others. In a comparable way, groups of Baboons have learned to take to the trees to escape from Lions but to come down and flee from the approach of men, even though years may have passed since men shot at them. Groups of different species of primates have also shown a marked conservatism in adherence to a home-range. The Howlers, Colobus Monkeys, Orangs and Gibbons spend almost their whole lives in the upper regions of the forest canopy. Viewed generally, the more arboreal the animal, the more restricted is its horizontal range. The Gibbon is estimated by Eimerl and De Vore (1966) to spend its entire life within the same one tenth of a square mile. A group of 17 Howler Monkeys was estimated to range over half a square mile while a group of 25 North Indian Langurs which are very much at home in the trees, but spend much of their time on the ground, ranged over 3 square miles. Mountain Gorillas were estimated to range over 10–15 square miles and Baboons a little further. Because vision is restricted by the foliage, arboreal primates are generally noisier than the more terrestrial species. Calls assist individual members of the group to maintain closer contact and, more important, give notice of the position of the group to other groups. Being in competition for many of the same foods, survival is best ensured if conflicts are avoided and an effective spacing obtained. The day begins for Howler Monkeys, for example, with a roaring session which tends to be repeated when groups come near to each other or move to a new feeding place. Conservatism is also evident in several species in the frequent use of familiar sleeping places.

The foregoing are some general features of the life of the primates. There follow accounts of studies with particular species. Not all species have been equally studied. Indeed it is difficult to study the social organisation of inhabitants of the forest canopy and considerations of space dictate a selection in the present work.

THE HOWLER MONKEYS

The Howler Monkeys (*Alouatta palliata*) are the most studied of the New World Monkeys, probably because of the special facilities which exist on Barro Colorado island which became an island in 1914 with the damming of the Chagres river in Panama and the formation of Gatun lake as part of the Panama canal. Since 1923, the island has been a protected reservation, but is open to scientists and a system of trails across the island facilitates study. Between 1932 and 1959 census studies (Carpenter, 1965) indicate an increase in numbers from 398 individuals in 23 groups to 814 individuals in 44 groups, giving an approximate distribution of adult males 16%, females without young 29%, females with young 16%, infants 17% and juveniles 22%. Females reach reproductive maturity at 4 to 5 years and males from 6 to 8 years.

Howler Monkeys are capable of moving effectively over the ground on the occasions when they come to earth but they are predominantly arboreal. Unlike the Sifakas, they do not indulge in long flying leaps between trees but make good use of their long prehensile tail, in descending and climbing, head first. They are herbivorous and frugiforous and, strangely, have a small atrophic thumb which appears incapable of attaining opposition with the other fingers of the hand. The grip is obtained by opposition between the second and remaining fingers. Fine manipulation with the hands is thus not possible and in opening nuts and fruit, they have been observed to drop an appreciable amount of the food they handle. Nevertheless, they appear to be well adjusted to their habitat which they share on Barro Colorado island, which is only six square miles in area, with the Capuchin Monkey (*Cebus capuchinus*), the Night Monkey (*Aotus trivirgatus*) and the Cottontop Marmoset (*Oedipomidas geoffroyi*) which are together at least as numerous as the Howlers. Unless relieved by cropping or disease, the density may eventually become excessive; but when reported by Carpenter (1965) and by Southwick (1963), the social organisation of the Howlers appeared adequate for the circumstances.

Groups of Howlers appear to have preferred areas and home-ranges which may overlap with that of other groups, but the areas are usually not occupied simultaneously. They do not defend specific boundaries but defend the area where they are, usually by howling battles, until the other group withdraws. When groups approach each other, it is apparent that all of the animals are disturbed. The males roar. The females whine. The young cease to play and the group as a whole becomes more compact. Again, each day begins with a tremendous roaring session by each group which attains a maximum volume, far in excess of that recorded at any time of the day, between 5 and 6 a.m. One result must be that all groups within a considerable distance know the approximate location of other groups within the forest canopy. An effective spacing and sharing of the food supply is thereby ensured and costly fighting is avoided. Southwick (1963) records that he has never seen a Howler Monkey with wounds, which can often be observed among the Rhesus Monkeys and Hamadryad Baboons.

Within the group, dominance among males tends to be expressed in the space accorded to the individual. There may be mild competition for females and consort relationships are formed successively, but once a relationship has been formed, the other males tend not to interfere for its duration. The greater part of the hostility within the group takes place between adult and near adult males and occasionally biting occurs. This may be the reason for some males leaving the group and for the existence of 'all-male' peripheral groups, a phenomenon noted in other primates. Their position relative to the group could rarely be in doubt because of the excellent vocal communication and it has been noted that they frequently follow a group —not necessarily their original group—for a time before they are accepted. Females are not recorded as forming extra groups. There was no evidence of the males of one group forming consort relationships with females of other groups, indeed membership of the same group appeared to be a necessary condition for reproduction. New groups may form as a result of a splitting-off or budding process, possibly as a result of increasing size and antipathies between prominent males. Where groups come together temporarily, they are not strangers and may have once been members of the same group. There is no evidence of strange groups coalescing.

Conflicts between female Howlers appear to be quickly resolved and may sometimes be prevented by the intervention of the males, who have not been observed to attack females. Like other New World Monkeys, the female Howlers do not reveal swelling of the genital regions in oestrus. At this time a female will approach a male and if he is not responsive, others will be approached. As in other primate species, many females of the group display interest in the newborn, both touching and nuzzling it and permitting it to climb on their bodies; but they have not been observed to carry it away. In these early weeks, interest in the newborn appears to be much greater than in the mother.

In common with many other mammalian and primate species, the young form groups for play activities which among the Howlers are particularly intense including chasing, wrestling, play-fighting and a considerable amount of 'ganging-up' on larger animals. As in many primate societies, incipient sexual components such as mountings occur, but without the distinct components observed among Rhesus Monkeys. There seems little doubt that in play in the early years, the young learn many basic features of the social order such as status relationships, forms of communication and the role of the sexes.

THE INDIAN LANGUR

The reports by Jay (1962, 1963, 1965) provide valuable insight into the social life of the Langur of Northern India (*Presbytis entellus*). The information derives from studies in widely separated areas. One of the chief studies was centred upon Orcha Village, Bastar district, Madhya Pradesh in Central India. This is a tribal area of minimum cultivation, low density of population and surrounded by reserved forests inhabited by other animals such as tiger, leopard, wild dog, wild pig, hyena, deer and small forest mammals. These groups were difficult to observe and could only be followed at a distance of about 50 feet or studied from semi-concealment. The other centre was near Kaukori Village, 14 miles from Lucknow in Uttah Pradesh. It is situated in the densely populated and cultivated Gangetic plain where the Langurs are more accustomed to the sight of man and observation was naturally easier. Like Rhesus Monkeys living in cities or near settlement and apt to be chased, these Langurs were more tense and aggressive. By wearing coloured clothing, behaving submissively and avoiding eye-engagement (a threat for many primates), Jay gradually came to be accepted and could study behaviour from within the group which numbered 54, larger than the forest groups which ranged from 10 to 28.

While adept in the trees, and reverting to them quickly in time of danger, Langurs spend much of the daylight hours on the ground, returning to the trees for the night. Their diet consists largely of mature leaves which are not highly nutritious. In consequence, they must consume a large volume which supplies them with enough moisture to be independent of sources of water for several weeks. Within the groups studied, external dangers and the pressure of numbers did not appear to be great and issues of dominance did not appear to be acute. Very few wounds were noted and the general manner of the group was relaxed. The numbers of males living apart from the bi-sexual, social group, a common feature among primates, was small. Dominance among the males can only be inferred from close observation. A dominant male is less active in minor disputes than subordinates. He is quietly given a larger area of personal space, and priority in matters of food, access to females in oestrus, and favourable positions. A dominant male will pause slightly on meeting another and the sub-dominant quietly moves aside. Animals of inferior rank are generally more observant of higher ranking animals and, as with other primates, presenting of the buttocks is an indication of submission. There are many subtle forms of communicating status, such as the duration of eye-engagement, grimaces, grunting, coughing, barking, and slight shifts of bodily posture. Among females, dominance may vary in different situations, with stages of the reproductive cycle and with the ranking of her male consort. Females may frequently combine in small groups to dominate another female or a sub-adult male. A mother with a newborn infant appears to withdraw temporarily from dominance interactions. Jay (1965) mentions the difference between the relaxed state of the Langur groups and the more boisterous tone of a group of Rhesus Macaques. On one occasion, a pair of Rhesus Macaques (male and female) were observed to be living within the Kaukori Langur group. They both assumed dominant roles within the group and on occasion the female could break up a squabble between male Langurs.

Of special interest among Langurs is the permissive behaviour of Langur mothers in respect of the newborn which clings to the mother and is shielded from other females by the mother's body. When dry, and on the first day of life, other females may touch the neonate and several may hold it and at times over the next few days may carry the infant several yards from the mother; but she can always retrieve it regardless of the dominance ranking of the holder. Males who may jostle or accidentally strike an infant, on the other hand, may be pursued by the mother. Infants are normally weaned at about 12 to 15 months. The process is distressing for the infant. At first the rejection by the mother is mild, but the infant is frequently persistent and the mother resorts to turning away, holding off the infant, alternatively allowing the infant to cling and then rejecting or, on rare occasions, striking the infant. The infants are nevertheless a strong cohesive factor in binding the females together and much grooming of the young is observed. The males display a marked indifference to the young until approached by them at about the age of 10 months. These furtive approaches culminate eventually in embracing the male, and among juveniles (15 months to 4 years) mounting of adult males, frequently several in succession, is common. Similar approaches are made to sub-adult males and rarely to sub-adult females although there is contact between the sexes of different ages in relaxed mutual grooming. There is a long apprenticeship in social relationships and sexual approaches. From 15 months to 4 years, play is the major activity of the young but thereafter, from 4 to 6 or 7 years, the sub-adult male, while a marginal participant in the main stream of group activity, has many contacts with males and females of the group and progresses through a dominance order with both males and females.

STUDIES WITH THE JAPANESE MACAQUES

The Japanese Macaques (*Macaca fuscata*) are the only wild species which inhabit Japan, ranging from a northern limit on Shimokita peninsula in Honshu to Yakushima island in the south. As indicated earlier, these monkeys behave with reference to a definite dominance order which typically becomes more intense in overcrowded conditions. On arriving at a feeding station, the central part of the procession consisting of the dominant adult males, adult females and infants occupies the feeding place. The younger males of the group (Oikia in Japanese studies) and a few solitary males occupy the periphery. They may enter the actual feeding place only when the central section has departed. Frequently, subordinate males are mounted outside of the feeding area, presumably as confirmation of dominance. A simple indication of dominance is obtained by throwing a piece of fruit between two males. The less dominant averts his eyes and does not touch it. Were he to do so, he would be attacked and probably mounted. The young males, after passing through the infant and play group stage, tend towards the periphery at about the age of 2 years and, with increasing age and maturity, gradually come back to the centre of the Oikia. Imanishi (1957) notes some evidence that the offspring of dominant females tend to be dominant over other young monkeys and that these more dominant youngsters are subsequently less likely to leave the group. The dominance of the mother is related to some extent to the dominance of her male consort and it is conceivable that these dominant offspring become more familiar with the dominant members and thus identify more strongly with the Oikia.

The Japanese Macaques are remarkable for a manifestation in certain Oikiae of 'paternal' care which is in marked contrast to that shown by the Langurs. Adult males of high rank in the dominance order in the Oikia of Takasakiyama (5 out of 6 leaders and 9 out of 10 sub-leaders), for example, were observed (Itani, 1962) to have taken particular infants in their first and second year of age under their care. This included caressing, grooming and protection during crises. The 'adoption' frequently takes place when the mother is occupied with or about to have a new infant. Of the 1-year-olds, identification of sex was possible in 62 cases of which 28 were male and 34 female. Of the 25 2-year-olds, 5 were male; but this does not necessarily indicate a sex preference, because the 2-year-old males tend to move to the periphery while the young females stay in the centre of the group where the leaders also remain. No relationship was observed between rank order and paternal care (leader class—mean 5·8, sub-leader class—mean 6·0); but a comparison of ranks in the delivery season of 1954 with that in 1960 revealed that those males whose position was most challengeable showed more paternal care. Ratings of the adult males involved showed that those high in sociability and interest in the central part of the group and low in aggressiveness revealed the highest amount of paternal care. Further, it was noted that an adult male protecting an infant usually became much milder in manner; but there were some exceptions which were noticeably more aggressive.

Another remarkable feature of the studies of Japanese Macaques is the evidence of acculturation or the acquisition of new dietary habits. According to Kawamura (1963), different Oikia vary in their menus. In the Minoo Ravine, the Macaques could remove the earth on the slopes to obtain the roots of trees while Itani is reported to have found no evidence of this in the Oikia of Takasakiyama. Another Oikia was found to invade rice paddies frequently while yet another, which for years had encountered paddies in their nomadic life, were not known to have damaged them. The one food which the monkeys can be relied upon to accept is milk and one suggestion is that progression to other foods may occur by the young, in close contact with the mother, picking up the droppings from her mouth. There is a marked difference between wild and caged animals. Those in captivity will often eat new foods immediately. Wild animals appear not to regard the items as food but the manner in which new foods eventually become accepted by wild Oikiae varies. The eating of candy was established in the Takasaki Oikia in about a year and a half among half of the troop. It began among some of the 2- to 3-year-olds and spread among this age group fairly rapidly. It then passed to the mothers in close contact with them and was passed from the mothers to their infants. The practice of washing sweet potatoes was observed to begin in another Oikia at Kosima by Kawamura (1963) with a 1½-year-old female and rapidly spread among her playmates. Her mother and some other mothers acquired the habit and it was transmitted to their babies but the family of the originator was the first in which all members acquired the habit. In this Oikia, the adult males did not indulge in paternal care of the young and none of them acquired the habit of washing the potatoes.

Acculturation usually begins among the younger members who

appear to be less rigid, but an exception has been recorded. Yamada working with Kawamura (1963) noted an adult male of the B Oikia of the Minoo Ravine which was observed to have acquired the practice of eating wheat. This was transmitted to the predominant male from whom it passed to the predominant female and through her to her family. A female of lower rank also imitated the dominant female and then passed the practice to her family. In four hours the practice spread through the whole troop, probably because of the reinforcement from the factor of dominance. Nevertheless, the variation in the receptivity to new foods and practices between different troops is considerable and challenges explanation. The acceptance of new foods may be part of the uneven pattern of partial domestication as increasing use of the land restricts the habitat of a hitherto wild species and they become increasingly dependent on provisions left for them by man.

In the wild, the home-range of the Oikia varies from about 1 square km to 8 square km, according to numbers and the availability of food. Contact between members is assisted by cries which range from contact calls, warning calls, defensive and aggressive calls to sounds peculiar to females in oestrus. Conflict between Oikiae is reduced by threat displays such as cries and shaking of branches when two troops meet. All told, the Japanese Macaques display many features noted by other primates. There is the internal status organisation of the group, the long period available for social learning in the play of the young, and the intensification of the dominance order with restriction of space.

STUDIES OF THE RHESUS MONKEYS

The Rhesus Monkey (*Macaca mulatta*) is a common inhabitant of northern India and was studied by Southwick et al. (1965) in several different settings. It was estimated that in the state of Uttar Pradesh the Rhesus population in 1959–60 was something in excess of 800,000 and of these, 46% inhabited villages and neighbourhoods, 17% towns, 13% cities, 12% forests, 9% roadsides and canal banks and 2% various temples. At the time of the study, 16% of the 280 villages sampled had resident monkeys which do very well in a habitat which provides shelter, food and water; but the tradition of tolerating monkeys in the villages was declining and total numbers and group sizes in villages were falling.

The main area of study was the Achal Tank area of Aligarh some 80 miles south-east of Delhi where among a group of 12 Hindu temples surrounding the tank or artificial lake, three groups of Rhesus, numbering approximately 51, 34, and 19 and 3 peripheral with some splitting of groups and interchanging, lived without molestation. As with many other primate groups, individuals were noticed to leave the group for short periods. The groups consisted of three main sub-groups, namely, a central group consisting of dominant adult males, adult females, infants and juveniles, a dominant male sub-group consisting of a dominant male, adult females, infants and juveniles, and a peripheral sub-group of young males of subordinate rank. These males were very aggressive and frequently initiated inter-group hostility. The movements of the group appeared to originate with and were largely influenced by the dominant males of the central group, though not from the front. The male dominance order appeared to be strict and was maintained with much agonistic behaviour, in contrast with the females who displayed the minimum of agonistic behaviour and indulged in much mutual grooming. Despite these differences, both males and females had particular 'friends' of their own sex with whom they often associated and combined in defence. Females groomed males at all periods, but males in the main groomed females in the consort period which lasted for about three days. Unlike the Japanese Macaques, there was little evidence of paternal behaviour and frequent attacks by males on juveniles were recorded. The relationship between mother and young endures for about a year and is not normally interrupted until a sibling is born. At about the age of three months, the infants leave the mother for periods of play with other infants, but are never very far from the mother. Behaviour in the playgroups is typical of many other primate playgroups including chasing, play-fighting and mounting.

Of the groups in the temple area, each had a particular resting place for the night and certain areas in which, during the day, they were dominant to judge by their weaker aggression towards the boundaries and withdrawals by other groups when within the area. Although these territories overlapped, the different groups usually avoided contact. One group, strangely enough, would withdraw from the temple area for a few days at a time; but when two groups came unexpectedly into contact, fights occurred, led by the males, with females and juveniles joining in and young, adult subordinate males playing a leading provocative role. The noise of fighting usually brought monkeys scurrying from all areas with much excitement and displays of hostility.

Another study of Rhesus Monkeys has been provided by Koford (1962) from observations of Rhesus groups on Cayo Santiago island off the east coast of Puerto Rico. To this small island of about 40 acres, 400 *Macaca mulatta* were brought from India in 1938. Despite irregular supplies of additional food, they reproduced well and several hundred were removed from the island up to the early 1950s. At the time of Koford's study, the population, from natural causes, was about 420. The diet of the animals consists of bark, twigs, leaves, fruits and flowers and, chiefly, pelleted monkey food supplied in five metal hog-feeders spaced 100 to 250 yards apart. Opportunities for study occurred when the monkeys came to the feeding stations. In 1958, the population was divided into two bands, one larger than the other and consisting of over 100 monkeys. Probably in the autumn of that year, this band divided and subsequent divisions gave six bands in 1962 with numbers 140, 140, 50, 30, 40 and 20. In the main, larger bands have priority but the status of the band leader is important and can outweigh mere numbers.

The general organisation of the band was comparable with that found at the Achal Tank. The central group of a few highest ranking males were most influential in group activities, and dominated the sub-dominant and peripheral males. The leader could be readily distinguished by his fine physical condition, assertive stride and high tail position. Peripheral males were generally subordinate to females and so, too, some sub-dominant males. In some groups, the highest ranking female was subordinate only to the leader and it was noted that the sons of high ranking females had a relatively rapid rise to the higher ranks, possibly because of greater familiarity with the dominant animals. Grooming is the most frequent contact between animals, about 80% of the total activity occurring between the mothers and their various young, including their mature daughters and their young, and between brothers and sisters. In the mating season, grooming between adult males and females is a prominent part of the relationship. Mounting accompanied by definite pelvic thrusts by males occurs throughout the year and usually follows the dominance order; but sometimes the dominant is mounted and among adolescent males there is on occasion a reversal of roles.

Leadership of a band usually endures for a year or longer depending upon circumstances, but one male held office for four years. Deposed leaders may either hold a lower rank in the group or on occasion become a sub-dominant in another group. Among peripheral males, changes of band membership is common and frequently occurs in the autumn when because of the breeding activities, aggression and restlessness are at a high level.

Unlike the Langurs, the Rhesus Monkeys appear to be relatively poorly equipped with mechanisms to avoid inter-group fighting. Southwick (1963) was particularly impressed with the ferocity of inter-group fighting among the Rhesus at Achal Tank and mentions the absence of ritualised patterns of vocalisation which would enable groups to avoid each other. Unfortunately, close studies of the

Rhesus in a forest environment without constriction are difficult; but it is of interest that in two widely separated environments, so many points of similarity in behaviour can be observed.

THE GELADA BABOONS

We are indebted to J. H. Crook (1966) for a study of the Gelada Baboon (*Theropithecus gelada*), a ground-dwelling, vegetarian primate living in the precipitous high country between 6,500 and 16,500 ft in Ethiopia. There is a long, dry season, and because of the altitude is temperate. The Gelada is subject to predation by leopards and foxes and has adopted the practice of grazing not very far from cliff faces to which the group returns at night or in time of danger. The basic unit of the Gelada group is the adult male, a group of females in all stages of oestrus, infants and juveniles; but when grazing is good the smaller groups come together and in Semyan province in February, a herd as large as 400 and an average size of 156 for 24 herds was recorded. In difficult conditions, the large herds disintegrate. Crook thinks that one-male groups are an adaptation to sparse feeding resources since a smaller proportion of the available food in a given area goes to those least involved with infant nutrition.

Browsing herds may be dispersed over a wide area but they are never more than a mile from a cliff face with the infants and juveniles nearest the cliff, interspersed with the mothers and some adult male leaders. Sub-adult males increase towards the outer flank until, on the extreme flank, only adult males occur. Movement in the main is leisurely. The animals rarely move before daylight when they come out into the sun and groom. Movement is usually initiated by the male section of the herd and is normally for a distance of about 300 yards when the initiator settles. Thereafter, others rarely pass him. There appear to be no leaders in this process. It is a gentle, surging movement which any one of the males may initiate. The Geladas revealed a fair degree of consistency in family membership, but as with other primate groups interchange between groups occur and the herd organisation is not such as to prevent this. There are also all-male groups consisting of old males and sub-adult males which may move independently and remain together for several weeks. The adaptation to the narrow band of grazing adjoining the cliffs is the Gelada's solution to the difficult problem of predation. The Chacma Baboon (*Papio cynocephalus*) and the Hamadryas Baboons (*Papio hamadryas*) in other parts of Africa may range more widely and may even form a working relationship with other animals such as Grant's Gazelle whose scenting ability supplements the vision of the Baboon in avoiding predation. It is of interest that there are no records of the systematic use of tools by the Baboons when sharp sticks or stones would be helpful in prising away soil and uncovering the roots of plants and grasses.

THE GORILLA

There are two sub-species of Gorilla—the so-called lowland Gorilla (*Gorilla gorilla gorilla*) which inhabits the lowland forests of southern Nigeria, Gabon, the Cameroons and the rain forest of the Congo Basin. More is known of the Mountain Gorilla (*Gorilla gorilla beringei*) from the studies of Schaller (1963) in the area of the Virunga volcanoes of Albert National Park, Congo. In this area, the Gorillas were observed to penetrate as high as 13,500 ft where the humid, equatorial forest gives way to open slopes and frost occurs at night. Contrary to all popular legend of the most terrible of monsters and film versions of King Kong, Schaller had little difficulty in approaching these gentle animals. If he encountered a group on a forest trail, the adult male would rise, give a roar of alarm and retire into the trees. Schaller noted the peculiar head-shaking gesture with which the animals sometimes greeted a stranger and when he copied this the gorillas would turn and retire. The huge animals are almost exclusively vegetarian and require enormous quantities of food which includes vines, leaves, bark, roots and fruit. Adults, while capable of climbing, tend to be mainly terrestrial since only the stronger trees and branches support their weight. On the ground they move quadrupedally although they can shuffle forward bipedally and can stand or squat while feeding. Few Gorillas retain the ischial callosities of the old world monkeys and they have little need of them. Nests for night or day resting are constructed either on the ground or in the lower parts of trees by breaking or bending branches inwards. The leading male was on four occasions observed to be the first to construct a nest and others in turn nested in his vicinity. At times, infants played at buildings nests which were never used since they slept with their mothers; but on occasion, older infants built their own nests adjacent to their mother's nest and occupied them. Much of the day is perforce spent in eating. A day begins between 6 and 8 a.m. The animals eat steadily for a few hours, slowing down towards the middle of the day when a rest is taken and resuming in the afternoon until dusk, when the process of bedding down for the night begins. They tend to eat selectively. Food is rarely transported. Schaller observed no systematic use of tools (twigs or stones) in obtaining food and no instances of drinking. The damp vegetation is apparently sufficient. With no predators, the life of the Gorilla would appear to include few challenges and this feature may be in part responsible for the slow tempo and relatively benign character of their social life. Development to the full adult status of the silver-backed male and a weight of between 300 and 450 lb requires some ten or twelve years.

Groups of Gorillas were found to range in numbers from 5 to 27 with a mean of 17 and in the following proportions: silver-backed males 10%, black-backed males 9%, females 37%, juveniles 17% and infants 27%. Schaller noted that 4 silver-backed males and 3 black-backed males were lone animals for the period of the study. Lone females, as with other primate groups, were never observed. Only the dominant male remained with the most observed group for the whole period of the study. Individual males were noticed to have left the group for a few days and later to have rejoined without disturbance. In terms of sign (dung, broken foliage, scent and sounds of the group) they were probably still in contact. In the daily movement, the group moved with cohesion within a diameter which rarely exceeded 200 ft and it was clear that the leader largely determined the movement. He would stand motionless, facing in a certain direction while the others drew near. After a time, the group moved forward in that direction and it was clear that the members of the group at all times were aware of the approximate position and the direction of the leader's movements. The dominance ranking was not violently asserted but was evident in the precedence given on narrow trails, favoured sitting places or dry and sheltered places.

Groups appeared to confine their movements to definite areas or home-ranges of the order of 10 to 15 square miles. These overlapped with the home-ranges of other groups but when groups came close, there was little strife. On one occasion, two groups remained in close proximity (35 ft) for three days and nested in close proximity. Finally, on the third day, they mingled for five minutes before parting. Two groups were once observed to occupy the same nest-site for the night before parting in the morning. Groups were also observed to behave permissively towards one group and intolerantly of another. One instance was recorded of the leader of one group charging towards the leader of another group and approaching until their brow ridges almost met. The two groups drew apart later on the same day.

Apart from the noises of eating and disturbances of bushes, Gorillas have vocal methods of keeping in contact and probably of communicating. Harsh, staccato grunts by silver-backed males would stop female gorillas from quarrelling and were a probable warning of danger. Soft grunting at low pitch indicated an individual's position in the foliage when near at hand and rapid notes at higher pitch probably helped to maintain group cohesion when the group was more widely scattered. Short, loud barks were noted of animals while quarrelling and, given in the presence of man, were a warning of possible danger. Harsh, short screams uttered by quarrelling females were also made in the presence of man and probably indicated fear and gave warning. These could become longer and higher in pitch and were given by females and juveniles in the presence of man. The full roar of the male served not only as a warning but was probably also a form of intimidation. Because of differences in vocal quality with age, the sounds would give not only the position of the vocaliser but also his or her emotional state and the age and approximate standing of the vocaliser—all useful information in dense and obscuring foliage.

Chest beating by male Gorillas may form part of a complete sequence of nine phases of which the several acts may also occur either individually or in combination with only some of the others. Only the silver-backed males perform the full sequence, and this infrequently. The sequence consists of some 2 to 40 clear hoots given while standing, symbolic feeding on a leaf plucked from nearby vegetation, rising to full height on two feet while throwing vegetation in all directions, beating the chest with open, slightly cupped hands (also thighs, abdomen and trees) while sometimes kicking with one leg, rapid running sideways bipedally then quadrupedally while mowing down vegetation with the arms and, finally, thumping the ground with one or sometimes both palms. The throwing of objects and uprooting of vegetation is probably a displacement activity and is not uncommon when the animals are disturbed. The combination of several frequently performed acts into a ritual of several phases may be a displacement activity arising from a conflict between the impulse to flee and the impulse to attack and the need for some release of tension. The instigating circumstances for such a performance, apart from occasions when it appeared to be unprovoked, were the presence of man, the presence of another group, or some form of display by a member of the group. The ritual is intimidatory. The sheer size, force and destructive power of a large animal moving quickly could not be otherwise, and release from tension is obtained. Moreover, the enormous destructive power of the animals, which could extinguish the species, is channelled into a series of ritualised acts. Within the dense rain forest, the maximum effect of the display is probably not attained but the displacement of excitement has clear advantages for the group.

Observation of intra-group life is difficult in dense rain forest; but Gorillas appear to be exceptional among primates in that relatively little grooming is evident. Females groom infants and, on occasion, one another; but no female was ever observed to groom the dominant male. Despite the dearth of grooming which is thought to strengthen the cohesion of a primate group, the Gorilla bands are comparatively stable. It may be that with no predators, the ability of individuals to leave and rejoin the group removes those in a state of tension at the crucial time and much prolonged disruption is avoided.

Another feature of the gorillas is that they are not, on the whole, playful and where instances of playful behaviour were observed by Schaller, the animals were infants or juveniles. Play in fact appeared to have ceased by the age of 6 years and almost half the observations of playful behaviour were of young animals playing alone. In social play, the proportion of infants and juveniles were roughly equal; but infants played alone about twice as much as juveniles and began to play vigorously at about the age of $4\frac{1}{2}$ to 5 months when they became capable of walking away from their mothers with some assurance. Although the juveniles were on occasion five times the weight of the infants and vigorous games such as 'king of the castle' and 'follow my leader' were interspersed with chasing and wrestling, the infants were never hurt. Adults including the silver-backed males were extremely tolerant of the youngsters' play and playful approaches, even when in uncomfortable positions; but only one instance was recorded of a female reciprocating. Some phases of the play are illustrated (Fig. 25).

In general, the life of the Mountain Gorillas appears to be marked by much less aggression and irritability than that of other primates. The relatively slow tempo of their lives may be in part due to their security and the necessity of taking large quantities of less than optimally nutritious food. The relative absence of provocation, on the other hand, is not conducive to behavioural modification and development.

THE CHIMPANZEES

The Chimpanzees are members of the genus *Pan*, which of all the great apes has the widest distribution. In Africa, representatives of the genus occur from the Atlantic coast to the upper Nile in a belt extending approximately 10° North and South of the Equator. The two major species are *Pan paniscus*, the pigmy chimpanzee which occurs on the southern side of the great bend of the Congo river, and *Pan troglodytes* which, in different sub-species and races, is more widely distributed. *Pan troglodytes schweinfurthii* has been extensively studied by Reynolds and Reynolds (1965) in the Budongo Forests of Uganda and in the Gombe stream reserve by Van Lawick-Goodall (1968).

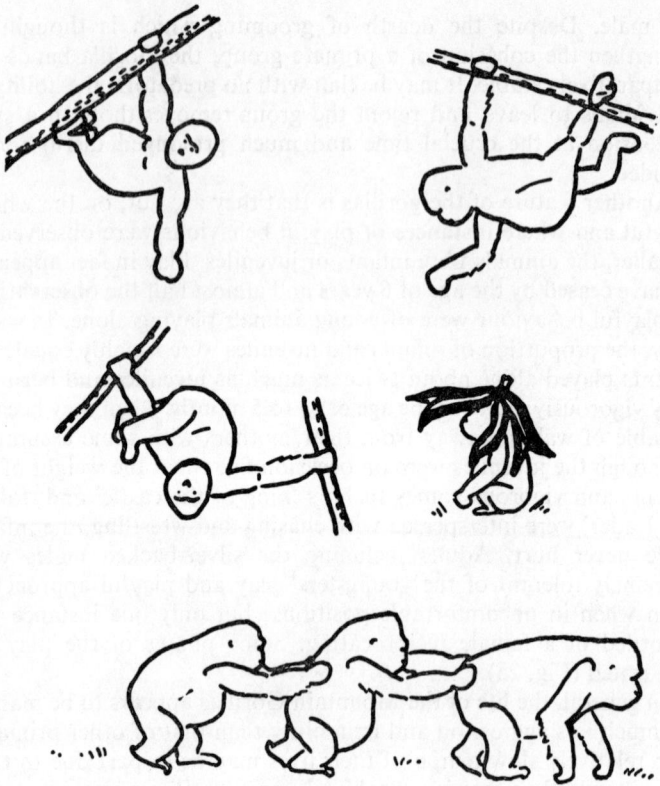

Fig. 25

The upper four pictures show phases of individual play by young Gorillas.
Lower right, an infant plays with a lobelia stalk. The lower picture shows a
phase of the less common group play—three juveniles join in a 'snake dance'.
(After Schaller, G. B., *The Mountain Gorilla*, Chicago Univ. Press, 1963)

These Chimpanzees are relatively large animals. The males average 110 lb and the females 90 lb. They are very active, noisy and adaptable. They can stand upright and walk for short distances; but their usual method of locomotion is quadrupedal, on the soles of the feet and the middle phalanges of flexed fingers. They can use holds on shrubs, trees and rocks to help them up steep slopes and snap off branches for braking effect in descending. They can also 'crutch' their way down steep slopes on the arms, using the buttocks as a brake. They have been observed to jump over streams and land like a broad jumper on two feet, or on all fours. They are adept in the trees and, although not as skilful as the lighter Gibbons, can brachiate quite well. They eat a wide range of foods including fruits, leaves, blossoms, seeds, stems, bark, resin, honey, insects, birds' eggs, and on occasion, the flesh and brains of small monkeys, baboons or antelope. Some reference to variations in the food habits of the Japanese Macaques (Kawamura, 1963) have been made and Schaller (1963) noted differences between different bands of Mountain Gorilla. Such differences are also found among Chimpanzees. In the Gombe stream reserve, the Chimpanzees readily ate plantation bananas which were made available to them, but in the Budongo forest, Reynolds (1965) noted that bananas placed on the trail were rejected or ignored.

Chimpanzees of the Gombe stream reserve were observed to carry food for some distances and to make and use tools—a criterion judged by the French philosopher, Henri Bergson (1911), to be evidence of the operation of intelligence. Leaves are crushed and dipped in water to assist in drinking and fine stems may be pared to a diameter which can be inserted into the holes to draw out ants or termites from their nests. These prepared instruments were sometimes carried for considerable distances. Chimpanzees also build nests, sometimes as high as 80 ft, but usually 30 to 40 ft up in the trees. Day nests for resting may also be constructed and those suffering from colds in the Gombe stream reserve were sometimes observed to build their nests and retire earlier. Unlike the Gorillas, Chimpanzees do not foul their nests. Nest-building apparently is learned by Chimpanzees. Wild Chimpanzees encaged after the age of 3 years will build nests several years later in captivity; but those separated from the mother in early infancy rarely do.

The composition of groups of Chimpanzees tends to vary widely from time to time from singletons to groups of 20, in respect of numbers, and also in composition. Van Lawick-Goodall (1968), from a total of 350 sitings, noted the following compositions: mixed groups (males, females and young)—30%, mature females with offspring—24%, mature and/or adolescent males with mature and/or adolescent females—18%, unisexual groups, i.e. all male or all female, mature and/or adolescent—10%, and lone individuals—18%. Compared with the Gorillas, the group composition is far more variable and yet Van Lawick-Goodall, over a period of time, was able to note a definite dominance hierarchy which was evident in subtle ways. Right of way would be conceded on pathway or branch or access to a favoured seating place. There was evidence, too, of borrowed precedence. A child of a high-ranking female might strike a female of lesser rank. Curiously, even normally dominant males would hold out the hand in a begging gesture for scraps to the animal who had killed a small Baboon, and on occasion a high-ranking animal, who would normally oust others in competition for bananas, would wait and watch if a lower-ranking animal was in possession and had already commenced to eat.

The attachment between mother and young is particularly close and prolonged among the Chimpanzees. The young leave the mother for the first time at the age of $3\frac{1}{2}$ to $5\frac{1}{2}$ months and are zealously guarded by the mothers from interference by others of the group. Transport for the first 6 to 9 months is usually in the ventral position, the infant clinging and being supported by one hand of the mother. As the weeks pass, less support is given. At 5 to 7 months, the infant begins to ride on the mother's back, at first lying flat and clinging with hands and feet and later in the jockey position, holding with the hands only. The close association with the mother, though gradually relaxed, continues up till the age of 5 or 6 years and there is evidence that the loss of the mother in the first few years seriously prejudices survival. In the Gombe stream study, two Chimpanzees who lost their mothers at about the age of 3 developed abnormal behaviour patterns. Weaning occurs at about $3\frac{1}{2}$ to $4\frac{1}{2}$ years and in some cases was probably expedited or initiated by the mother. One mother for example was observed to hold her hands across her breasts when the juvenile attempted to suckle but compared with the practice of the baboons and langurs, the process is benign. After the first few weeks of life, the infant Chimpanzees are groomed by the mothers and at about 6 months, in turn groom the mothers. As they grow older, social grooming with others outside of the family circle increases.

In the Gombe stream study, a system of auditory and visual communication was noted. Soft grunts indicated position and relaxation. 'Pant-hoots' greeted the approach of another or were a response to distant calls. A soft bark threatened a subordinate. The 'waa-bark' threatened a superior, usually from a distance. High-pitched screams were used when threatening superiors or members of other species of which the individual was afraid. The 'hoo' call is in a sense a warning and is used when a strange object or sound is encountered. There are a number of gestures and postures which were found to have a certain consistency. A slight upward tilt of the head accompanied by a soft bark was a threat at low or moderate intensity. Raised arms, and hands slapping downward or flailing outwards towards the threatened individual, were used in threat. A curious bipedal swagger, at times with little forward movement, appeared both in threat and courtship procedures. Van Lawick-Goodall also recorded a form of ritual 'rain dance' which appeared to be instigated by rainstorms. As the rain began, a group of Chimpanzees came down from the trees which were situated half way up a slope and began walking uphill. At the top, the females and juveniles climbed high into the trees but as the rain increased in force and the thunder crashed, a male charged diagonally down the slope, slapping the ground. This seemed to act as a signal and two other males followed, charging violently downhill, breaking and uprooting vegetation, leaping into trees and breaking off branches and finally running up trees at the bottom of the slope to arrest their momentum. After a brief rest in the trees, they would plod up the hill and repeat the sequence. The whole performance would cease just as suddenly as it began. It is difficult to decide whether this is a form of displacement activity associated with the excitement or fear of the storm or a gesture of irritation or perhaps even a form of play.

A number of facial expressions were also observed to be associated consistently with particular types of behaviour. The 'play-face' which accompanied soft panting sounds (laughing) exposed all the forward teeth and was usually an invitation to play, used when one juvenile approached another. It was also evident during strenuous wrestling or tickling. Glaring or staring fixedly at another with lips compressed was sometimes the prelude to attack or copulation. The 'pant-face' with lips pursed and pushed forward and eyes directed at another was a form of entreaty accompanied by the 'hoo' whimper, and used by infants seeking the mother or the nipple and by older animals when begging. It was quite evident from the Gombe stream study that anyone who had observed the Chimpanzees for a sufficient length of time could discern a pattern in the many forms of communication which was consistent with the general behaviour and responses of the animals.

Play among Chimpanzees is marked by great variety and includes wrestling, tickling, chasing, tug of war and chasing and struggling for objects such as a nut or item of fruit. Adults were often observed to enter into play as a result of pestering by infants and juveniles. Play between adolescent males could often become rough, resulting in one member screaming in retreat. Of 539 play sessions observed by Van Lawick-Goodall, 75% involved pairs of Chimpanzees, 19% involved trios and only 4% involved 4 or more. Frequency of play reaches a peak in the years 2 to 4 and, in general, declines as the animals become older. As in many other primate species, there is a strong sexual element in infantile play. Male infants by 6 months show interest in female genitals and frequently inspect them, and within the first year attempt copulation with receptive females. Frequent mountings of male and female infants by males are observed in play sessions.

Perhaps one of the most interesting observations from the studies of Chimpanzees is their use of weapons. The late Professor K. R. L. Hall observed that several species of apes and monkeys throw down branches at other animals when in hostile mood. Van Lawick-

Goodall has also observed Chimpanzees to throw rocks both over-
arm and underarm at Baboons who were disputing access to food.
Kortlandt and Kooij (1966) and Kortlandt and Van Zon (1967) have
also conducted experiments with Chimpanzees in the wild by exposing a model leopard, which moved the head, in the path of a group
of savanna-dwelling Chimpanzees. On first observing the 'leopard'
the group scattered, the females seizing the young and retreating to a
distance. The older males, however, soon advanced and began
throwing branches and sticks, spear fashion, several of which struck
the model leopard quite heavily. Eventually they advanced to prod
the model and strike it with sticks, hands and feet. It was interesting
that the film, shown at the International Ethological Conference,
1967, showed the first male to advance to close range turning to
shake hands with his nearest collaborators before advancing alone.
Forest-dwelling Chimpanzees on the other hand performed very
poorly in the same experimental situation. They were less consistently bi-pedal, showed virtually no concerted action and only a
vestigial use of weapons. One forest-dwelling chimp actually broke
off a branch and stripped it to the form of a useful spear and then
dropped it in front of the model leopard. Kortlandt suggests that
life in the semi-open woodlands and savannas favours the emergence
of protohominid and humanoid types of behaviour including bipedal transportation of food, social cooperation and the use of crude
weapons which eventually develop to spears which can kill at a
distance. He hypothesises that when spear-using hominids invaded
the central and West African savannas, the apes were forced to
retreat into the dense forests where their humanoid behaviour
declined and in this way exemplified the 'de-humanisation hypothesis'. Whatever may have taken place in the past development of
the Chimpanzees, their behaviour observed in recent years indicates
the emergence of tool-using, locomotion and communication which
are parallel developments in man.

MAN'S EARLY ANCESTORS

It is possible only to speculate as to how man's ancestors changed
from using their teeth to using weapons and evolved a social organisation. Climatic changes induce changes in habitat, and survival is
related to adaptation. The skills of the dense rain forest are not those
of the savannas and open woodlands where the emphasis is upon
freedom of movement over distances, vision extending over distances
(many chimps have been observed to stand up to gaze over tall
grass), and the ability to carry tools and even food. In open woodland,
away from moist vegetation, movement and subsequent settlement is

linked with water supply, until vessels can be shaped. In difficult times, survival would tend to favour the small family unit, and large aggregations, typical of periods of plenty, would tend to disperse as noted of the nomadic bands of the !Kung bushmen. Wider-ranging, all-male bands (as Crook, 1966, noted of the Gelada Baboon) would tend to withdraw pressure from the conventionally used areas leaving more for the family units which are reproducing and rearing young. As hunting became a possibility, some females would have to stay with the young and groups of males would be more effective in hunting and in defence against large predators, if they cooperated.

The social life of man's prehuman ancestors can at best be only a matter of speculation. The australopithecines or Southern Apes are thought to have existed at the beginning of the Pleistocene period over one and a half million years ago in south and east Africa. Finds of primitive stone tools are not sufficiently consistent to prove that they regularly used or made such tools. There is evidence from fossil bones found by Dr L. S. B. Leakey in the Olduvai Gorge in Tanzania in a layer dated at about 1,750,000 years ago of a possible new species termed *Homo habilis* (skilful or able man) which, from the form of the thumb, leg and toe bones appears to have been capable of standing erect and walking bipedally and gripping objects or tools. Controversy continues as to whether this species is within the range of variation of the australopithecines or should be classified as another species. The estimated volume of the skull at 670 cc is larger than that of any known australopithecine. The more rounded shape of the jaw and the dentition suggest a progression towards hominid features, and simple stone tools found in association with the remains suggest that *Homo habilis* probably was a tool maker.

From discoveries of fossil bones dating from the Middle Pleistocene period, about half a million years ago, in Java, Africa and China, emerges an impression of what was once called *Pithecanthropus erectus* (erect ape-man), now termed *Homo erectus* (erect man). The skull capacity is larger than that of *Homo habilis*. From fragments found in a cave at Choukoutien near Peking in China, a skull capacity of 900–1,200 cc is indicated and the stone tools are an advance on those found with *Homo habilis* at Olduvai. Ashes extending downwards through the layers in the cave for several yards indicate that Peking man used fire. Another notable find at Ternifine in Algeria, dating from the Middle Pleistocene period, provides a picture of a skull consistent with the classification of *Homo erectus* in association with stone tools, scrapers and primitive hand axes. The bones of large mammals found on the Peking site and the nature of the ash heaps suggest that this form of *Homo erectus* knew how to kindle and maintain fire, probably for protection and warmth, and may

Fig. 26

(a) *Australopithecus* sp., (b) *Homo habilis,* (c) Peking man, (d) *Homo sapiens.*

have cooked the meat from the animals, which would have been difficult for an individual to capture or kill and suggests cooperation at the level of the family or social group.

From the Upper Pleistocene period of about 100,000 years ago, there is evidence of further advance to *Homo sapiens neanderthalensis* (Neanderthal man) and *Homo sapiens sapiens* (modern man). The term Neanderthal derives from Neander near Düsseldorf, Germany, where skeletal remains were found and is applied to comparable skeletons which have been found in France, Italy, Belgium, at Monte Circeo in Italy and Teschik-Tasch in Uzbekistan. Associated with these skeletons are those of mammals of the last glaciation to extend over Europe, namely, woolly rhinoceros, mammoth, reindeer, arctic marmot, wolf and bison. The Neanderthalers were erect, bipedal, had a skull capacity of the order of 1,000 to 1,600 cc, made stone tools, hunted and made use of caves and fire. They also buried at least some of their dead and from the circles of objects found at Teschik-Tasch and Monte Circeo, there is suggestive evidence of some kind of ritual. The Neanderthalers appear to have vanished, or may have interbred or been displaced by Cro-Magnon man, a classification deriving from five partial skeletons and some infant bones found at Les Eyzies, in the Dordogne, France. Comparable remains have been found at Grimaldi in the Riviera, the upper cave at Choukoutien (Peking), the Niah cave (Borneo), the Paviland cave in Wales, Boskop in the Transvaal and several other sites. From the muscle markings on the bones and length of the bones, Cro-Magnon man appears to have been tall, fully bipedal and erect.

The successors to Cro-Magnon man were Modern man (*Homo sapiens sapiens*), definitely erect and bipedal with full opposition between thumb and fingers enabling him to grasp both thick objects such as the handle of an axe with the hand or a fine instrument between the fingers. Culturally, the significant developments by about 8000 B.C., as indicated by finds in the Middle East, were the decreasing dependence upon hunting, the domestication of animals and the rise of agriculture with consequent influence upon dispersions and the growth of larger, more concentrated communities where some cooperation was necessary and divisions of labour were desirable.

F

9

RETROSPECT

Some fundamental aspects of behaviour in which both individuals and groups of many different species are involved have been reviewed in the foregoing pages and, as suggested in the introduction, the point has been reached when in respect of these aspects of behaviour, the performance of different species can be compared with that of man.

With regard to the processes of attachment of mother and young, it is clear that effective attachment is essential not only for the survival of the young but also for normal, social development to the point where the young are themselves capable of becoming effective parents. In several species an important contribution to this development derives from extended periods of life within a stable group or community with ample opportunities for play by the young. Failure or rupture of the attachment in many wild animal species may be fatal for the young on purely nutritional grounds. Human societies are more skilful in organisations to ensure the physical survival of motherless or unwanted children; but the social costs of sustaining children deprived of adequate early attachment are likely to increase with the greater incidence of any features of the society which disrupt family life or deprive the young of the opportunities of attaining a conviction of solicitude. Granted freedom from interruption and adequate space for dispersion, the possibility could be considered that many animal communities may achieve greater success with all the problems of attachment.

In reflections on the proportion of young who attain normal, reproductive status and full social competence and effectively contribute to the defence and integrity of the group, it is important to consider the relatively large number of guide lines for behaviour which exist in different animal communities. The approach of the sexes and the individual males, and to a lesser extent the females who may participate, are in many species governed by a strict dominance hierarchy and in some cases an elaborate sequence or ritual. A high degree of conformity on the part of the individual is demanded but assuming adequate opportunities for dispersion, the probability of continuation of the group is enhanced. The full biological significance of the frequent observation of all-male groups among the ungulates and primates and individual male primates which withdraw from the main group can only be a topic for speculation; but among the possible results could be included a reduction of tension and an interruption of both reproductive activities and the demand on the nutritional resources of the habitat. Both results would be conducive to the continuing survival of the group and of the species. An interesting result would be the proportion of the all-male groups which are permanent 'drop-outs' who never re-enter the activities of the group. One important difference in favour of the animal communities would appear to be that the 'drop-outs' are less of a charge on the resources available to the community.

On the other hand, the strictures upon the individual, as in all autocratic communities, reduce for the group the element of challenge and the possibilities of modification. The greater social and cultural advances of human communities derive in part from the capacity of human groups and societies to tolerate the challenges of individuality. In this process, the greater powers of communication and periods when the issue of survival recedes are obviously important. The possibilities for cultural advance in settled communities would appear to be greater than for nomadic groups and, as factors conducive to the growth of settled communities, the growth of agriculture and technology must be regarded as important.

As indicated in the foregoing pages, individuals of many species are rarely without some form of reference to others of their own and frequently of other species because of the many forms of visual, auditory and olfactory communication; and within any one species, the means of influencing the behaviour of others by various means of communication are considerable. Warning, threat, distress, appeasement and soliciting signals are all basically activities for influencing the behaviour of others. They precede the emergence of articulate language in evolution, adding support to the view that the primary function of language, when it does appear, is not so much a means of self-expression as a means of influencing the behaviour of others. In all phases of life relating to dispersion, survival and the control of aggression, effective means of communication have evolved in animal communities, but clearly human communities, because of language, are far better able to pass on to successive generations the results of experience as a basis for further progress.

In the control of actual aggression within the species, however, the achievements of animal communities which are adequately dispersed would appear to compare very favourably with those of human communities. The rituals for the approach of the sexes and for combat between males, as well as the several appeasement gestures, would have the net overall effect of reduction of damaging or fatal aggression and such processes of restraint are the more effective because the performance of the ritual appears to be binding upon the behaviour of the individual with few examples of inconsistency or 'duplicity'. Predation upon other species is for the great majority of animals controlled by nutritional needs. Few animals, like the Dingo (*Canis dingo*) in Australia, kill large numbers of sheep, for example, grossly in excess of their needs in what could be likened to an orgy of killing. Like the destruction of a whole community of domestic fowl by the raiding Fox, there may be special and disturbing features in the situation involving concentrations of defenceless domestic animals which would only recently have been encountered in the evolutionary history of the predator.

One recurring observation of many animal species is the considerable amount of effort expended in marking the boundaries and in resisting encroachment upon territories. Another is the greater apparent confidence shown by an animal in its own territory and the examples of reduced confidence or assertion by animals outside their own territory. Yet another is the frequent association of the gaining and holding of territories with reproductive activities. The obvious result from these features of behaviour is greater dispersion; but the direct implication is that members of many species have a preference for some degree of 'privacy' or priority in the use of an area of space from which they can exclude others. Conversely, many experiments have shown that crowding is associated with a high incidence of neurotic signs and increased aggression.

The need for some measure of personal space and the propensity to resist encroachment upon specific areas of territory is a feature of behaviour common to a wide range of species. What could be described as an awareness of occupancy and ownership and even a respect for ownership is definitely not peculiar to mankind. The crucial issue is whether or not the several aspects of territoriality, which have clearly been of considerable importance in ecology and the evolution of behaviour, can be effectively set aside and without loss to the behavioural resources of the community as in planned societies where the private ownership of property is restricted. Another consideration is that territoriality in animal species ensures dispersion so that the environment is not exploited beyond its regenerative capacity and the fittest members of the species reproduce. The private ownership of property in human societies has not always ensured these results. In the same context, it should also be remembered that modern immunology and therapies are keeping alive large numbers of people from birth onwards. Corresponding cases in former generations would not have survived. Among animal communities, in natural circumstances, selection is against survival and reproduction by the unfit.

The human community like the animal community is an organisation providing conventional competition or the arena where the aspirations of individuals impinge, are resolved and concessions won or granted; but for communities with spoken or written language and thus with cumulative experience and greater potential for change, the problem is one of admitting challenges and changes without the costly destruction or dissolution of social organisations. So many revolutions in human society have been literally revolutions, namely, the destruction of an autocratic regime in the spirit of protest, to be followed before long by an equally or more repressive regime. The basic conservative elements in animal communities, of strict hierarchical orders, rituals and dispersion related to food supply can clearly not have the same influence in modern, human societies where technology, agriculture and trade have made possible large aggregations of population having no direct relationship to the local sources of food or food production, larger numbers in specific occupations, an egalitarian tradition and greater social mobility. The dispersion of industries and population must pass from genetically determined patterns of behaviour to informed and purposeful planning.

The various methods of finding the way about which have evolved in animal species indicate the degree of specialisation which evolutionary selection can ensure. It is remarkable that evidence of solar and stellar navigation, echo-location and underwater sonar-location should be found at levels of evolutionary development far earlier than man. The errors made by army and air force recruits give some idea of the extent of this achievement; but it is noteworthy that when man began to make long journeys across or under the seas, in the air or by land, methods of navigation, based upon principles which had already emerged in evolutionary development, were evolved by purposeful application to a problem of which men by personal and reported experience had been made aware.

Perhaps this is an appropriate point to return to the problems adumbrated in the opening chapters of the present work where some of the problems of explaining behaviour were discussed.

Having surveyed the performance of some species in certain phases of their behaviour which are essential for survival, consideration can now be given to the issue of where in evolutionary development conscious, purposive behaviour may be claimed to operate. At the human level, while some involuntary acts can be admitted, the presumption is made in language and in jurisprudence that there is a large measure of communality in the behaviour of different individuals, that feelings of disappointment, resentment, sadness, depression, elation and the intentions accompanying the behaviour frequently associated with these states of feeling, have something in common. This presumption is the basis of all 'verstehende' methods which seek to understand and in this way make clear or explain the behaviour of other human beings. Because of this presumption and the memory of how one felt in comparable circumstances, clinical workers feel justified that they have some understanding and can to a great extent explain the behaviour of a hitherto well-adjusted lad, for instance, who learns at the age of thirteen that his much loved and respected parents are in fact not his natural parents and straightway embarks on a period of delinquency and violence. In this case, there is awareness of and retention of the precipitating circumstances, ample variability of sequences of behaviour to achieve his protest and much independence of action, frequently against resistance. With other cases in which protest and resentment are manifested in delinquency, the instigating circumstances may not be known to the individual but he retains awareness of many of his former acts, can vary the sequence of his acts to achieve the same result and can behave independently of conflicting or opposing circumstances. His behaviour meets the criteria of purposive behaviour even though the more distant basis of his attitudes is unknown to him. At other levels of evolutionary development, application of the same criteria can be considered.

In the attachment studies of Sheep, Goats and Moose, the behaviour of the newborn young reveals much trial and error. There is a variety of movements until a result consistent with the previous history of the species is attained. The young repeatedly butt and suck at various points of the mother and other objects until milk is obtained. Deprived young will attempt to suckle upon other females in spite of forceful rejections. The fact that they learn where to suck and rapidly reduce the time of successive approaches to the teat of a mother at different positions in the field argues for powers of retention. The ewes, does and cows in sniffing and rejecting several young and accepting their own give ample evidence of retention of an impression and the ability to obtain a matching judgement. In their frequent search behaviour and evasion of capture to be with their young, ample variability and independence of action are demonstrated.

Attachment in birds of the type known as imprinting presents greater problems. If the imprinting is of the rapidly effected 'mosaic' type, the young bird cannot show great independence of action and variability of behaviour in acquiring the imprint; but some discrimination is nevertheless required. Nidicolous birds whose choice of sexual partners when adult is that of the species of the foster-parents, show evidence of retention of an impression and the ability to achieve a matching judgement. The Zebra Finches whose choice of sexual partners is restricted in this way nevertheless in other activities, such as nest-building and avoiding attack, show evidence of variation in achieving a particular result and independence of action. With those birds which acquire the attachment in the regulative way, i.e. presumably by following the imprinting object and learning its features to a degree which will mediate discrimination, there is again evidence of retention and since these birds in many tests show ability to generalise and follow objects of the same type or more attractive examples of the same type, some claim for independence and variability of action could be made.

In the methods of finding the way about which have been reviewed in the foregoing sections, perhaps the claims for purposive action by the Eels and Salmon need the greatest amount of additional support. The upstream migrations of Salmon against violent and changing pressures, and with the frequent necessity to test alternative routes up or around waterfalls, provide ample evidence of variability of the sequence of movements to attain a result consistent with the previous history of the species, and independence of action; but the evidence for retention in connection with the 'parent-stream' hypothesis could be much stronger. The very small proportion of recoveries does not indicate in many studies a greater than chance return to the parent stream although in other activities, such as learning to return to feeding points, fish have shown evidence of retention.

In respect of the evidence for navigation and communication by bees there can be little doubt that they meet all of the three criteria of retention, independence of action and variability of the behaviour in different circumstances to attain a familiar result. The feats of navigation and homing by birds must surely meet the same criteria and so, too, the whole mode of life of the echo-locating bats. The evidence for retention could be stronger and could probably be confirmed by experiments, but some powers of retention are surely indicated by the practice of the bats in returning to the same roosting places.

In the studies of the scattered community (Chapter 5) the ability of individual members to maintain an independent life for long periods, and yet one which is adjusted to the lives of others of the community and provides for periodical association, must be mediated by powers of retention, variability of behaviour and independence of action. In the more compact communities, the establishment and maintenance of even mild dominance orders, quite apart from other activities, calls for continuing vigilance and adjustment to the behaviour of others. It is difficult to conceive how this definite but flexible order could be maintained without some powers of retention, of interdependence of behaviour of individuals, independence of action by individuals and the performance of behavioural sequences in changing social contacts and physical situations.

Similar considerations apply to the various rituals in sexual approaches, particularly to phenomena of the 'lek' type. The individual, both male and female, reveals a sequence of acts indicative of choice between conflicting influences and opposition to external pressures. The repeated assertion of hard-won rights and the continuing interplay of successive phases of the rituals of each sex and rivals of the same sex in the maintenance of territory provide adequate evidence of variability and retention. In other manifestations of territorial behaviour such as boundary marking, the deference frequently shown to the resident, the greater assertiveness of the resident within its own territory, and the driving away of intruders, there is again evidence of retention, variability of behaviour and independence of action.

To the several studies of the primates, similar considerations would apply. The maintenance of even the mildest of hierarchical orders of dominance requires a continuing interplay of behaviour and reactions to the behaviour of others, mediated not only by gestures and vocalisation but the more economical and possibly more effective eye-engagement. To this must be added the frequently noted co-operation between species such as the Baboons with Impalas and Grant's Gazelles in maintaining vigilance and defence against predators and the fear of man manifested by a group of Baboons after one shooting incident. There are, too, among the primates, in common with several other species, the play groups of the young which appear to arise spontaneously and possibly depend upon some awareness of kind, of smaller, inadequate beings among others larger and more powerful, and kindred disposition in other young to play. The construction of nests and the use of tools, such as rocks, crude clubs and rough spears and 'probes' for ants' nests by Chimpanzees are indicative not only of a capacity for independence and variability of action, but also of behaviour related to situations not directly present. In turning to seize or in fashioning a rough tool, the animal does indicate an awareness of a relation between the tool or weapon and the situation in which it is used.

From the foregoing considerations, it will be apparent that the criteria of purposive behaviour can be applied to the behaviour of a wide range of animal species, many of which are representative of stages of evolution much earlier than man. In meeting the criteria of purposive behaviour, namely, independence, retention and variability, it is difficult to dissociate purpose from a presumption of awareness. To behave independently of environmental forces and to achieve a result consistent with the behavioural history of the species in different and changing circumstances, suggests some form of awareness of features of the environment, at least of their position in relation to the animal's own body and, on occasion, an awareness of internal bodily conditions such as conditions of excess or deficit and some of the environmental features relevant to the abatement of these conditions. If one observes predatory behaviour in many species or attempts to capture a runaway horse or a variety of wild and domestic animals, one is bound to admit that the animal's appraisal of environmental features coincides closely with one's own and that many movements associated with one's intentions are reacted to and at times even anticipated. None of this behaviour would be possible without some element of retention which involves the ability to relate or associate perceived features and to retain the association. It is arguable that there are differences in the associations of animals and men, that a given external situation will frequently give rise to a greater range of meaning and foresight as to possible uses because of the greater range of knowledge available for association; but it is difficult to argue that the awareness of a given situation and the meaning of it for many animals is so different as to be qualitatively discontinuous with that of man or again to indicate a stage in evolutionary development at which the discontinuity would occur.

Perhaps an early stage in the development of the power to associate impressions of the different features of the environment and to retain them is indicated by the experiments of Yerkes (1912). A very large number of trials was necessary before the worms learned to turn to one side at the crossroads of the maze; but there was some evidence of retention. The learning was very slowly acquired and the relationship, if used at all in further learning, would almost certainly be slowly applied; but there is evidence of the formation of an association and of its retention. To claim the existence of a form of awareness in this case may claim too much but it is submitted that from some such processes of association or relating, awareness emerges.

If the criteria of purposive behaviour are evident in the behaviour of many animal species, it is defensible to use many words consistent with the operation of purpose, such as 'chase', 'pursue', 'strive', 'intercept', where the criteria are met, and it will be apparent that for the behaviour of many animal species, the approach represented by Tolman's (1932) purposive behaviourism is defensible in principle. The difficulties are the considerable amount of experimental work and statistical validation which would be necessary before the activities of the life cycle of a species could be described in Tolman's terminology. But putting these practical difficulties aside, the scope for purposive behaviourism is greater in 1971 than it was in 1932. Studies by ethologists have shown that many phases of the behaviour, even particular gestures, are specific to the species and follow a demonstrable pattern of inheritance. The neurological correlates of these heritable features have in several cases been demonstrated so that a considerable proportion of the animal's behavioural movements may be said to be prescribed by known and heritable neurological mechanisms and the role of these movements and gestures in communication, threat, appeasement or sexual approach is now more widely studied and better known.

Assuming, as Tolman did, that the animal's behaviour is purposive, we are now in a better position to indicate the behavioural resources available to the animal and their social significance. From close study of some life cycles, we can presume to indicate what the animal is trying to do, how it does it and the effects on other animals. In great part, this improved position is due to the work of the ethologists who have from the beginning studied animal behaviour objectively and by implication accepted behaviour as purposive. A threat display or sexual display is described in the context of an animal purposefully seeking to defend its territory or rank or to acquire a mate. The fact that the display fits well into an ethogram of related species and has demonstrable neurological correlates does not disturb the overall pattern of explanation. The animal is often purposively seeking the situations or stimuli to release or satisfy innate propensities and patterns of behaviour.

The position with regard to the explanation of human behaviour is more difficult, first on the grounds of sheer complexity. Historical, sociological, genetic and situational factors are known to be relevant. Unlike animal species, gestures with human beings have not the same reliability as indicators of a person's intentions. Because of the immense range of learning and cultural differences, a given situation probably does not convey precisely the same meaning to any two persons and the person's conception of himself in relation to society, which can frequently only be surmised, may be an important determinant of behaviour. As indicated in Chapter 2, there are several methodological problems which no one system has yet surmounted. A purely ethological approach would yield a coordinated body of data, but with so much of the possible determinants of human behaviour not discoverable by outwardly observable signs, important gaps in the system would appear. One difficulty is that being more concerned with human behaviour, in education, delinquency, mental health, industry, social and political life, we need adequate explanations more urgently in order to deal scientifically and humanely with the several problems which arise. Despite the methodological difficulties, an underlying rationale is presumed and we proceed with an amalgam of methodologies. Not unnaturally, the system, which as a hypothetical system coordinates the greatest range of animal and human behaviour, is that of the late Professor Wm McDougall. It was uncompromisingly based upon the assumption of his particular form of vitalism, that 'wherever there is life there is mind', that all animal and human behaviour was purposive and 'shot through' with intelligence. The more consistent features were apparent through the operation of a number of instincts, each in its operation mediated by a 'neurological correlate' which in 1908 had only hypothetical status but which now appears more defensible. Its basic concepts had much in common with the ethological approach, and from ethological studies the system has not unexpectedly received greatest support.

REFERENCES

AGAR, W. E. (1943). *A Contribution to the Theory of the Living Organism.* Melbourne Univ. Press.

AHRENS, R. (1954). *Zeits. exp. und. ang. Psychol.,* **2,** 402, 599.

ALEXANDER, G., & WILLIAMS, D. (1964). *Science,* **146,** 665.

——(1966). *Anim. Behav.,* **14,** 166.

ALLEE, W. C. (1952). *Colloq. int. Cent. nat. Rech. sci.,* **34,** 157.

ALLPORT, G. W. (1947). *Personality: A Psychological Interpretation.* Constable, London.

ALPERS, A. (1960). *A Book of Dolphins.* Murray, London.

ALTMANN, M. (1952). *Behaviour,* **4,** 116.

——(1958). *Anim. Behav.,* **6,** 155.

——(1963). In Rheingold, H. L. (Ed.), *Maternal Behaviour in Mammals.* Wiley, N.Y.

AMBROSE, J. A. (1960). Unpublished thesis, Ph.D. London.

ARDREY, R. (1967). *The Territorial Imperative.* Collins, London.

BACKHAUS, D. (1960). *Zeits. f. Tierpsychol.,* **17,** 345–50.

BAERENDS, G. P. (1941). *Tijdschrift voor Entomol.,* Deel, **84,** 68.

BAERENDS, G. P., & BAERENDS-VAN ROON, J. M. (1950). *Behaviour,* Supp. 1, 1.

BEATTY, H. (1951). *J. Mammal.,* **32,** 118.

BELLROSE, F. C. (1958). *Wilson Bull.,* **70,** 20.

——(1963). *Auk.* **80,** 257.

BERGSON, A. (1911). *Creative Evolution* (trans.). Macmillan, London.

BIERENS DE HAAN, J. A. (1948). *Behaviour,* **1,** 71–80.

BILLINGS, S. M. (1968). *Auk,* **85,** 36.

BLAUVELT, H. (1955). In Schaffner, H. R. (Ed.), *Group Processes.* Josiah Macey Foundation, N.Y.

BLOESCH, M. (1956). *Orn. Beob.*, **53**, 97.

——(1960). *Orn. Beob.*, **57**, 214.

BOURLIÈRE, F. (1964). In Howell, F. C., & Bourlière, F. (Eds.), *African Ecology and Human Evolution*. Methuen, London.

BOWLBY, J. (1944). *Int. J. Psycho. Anal.*, **25**, 107.

——(1951). *Maternal Care and Mental Health*. WHO, Geneva.

BOWLBY, J., AINSWORTH, M. B., BOSTON, M., & ROSENBLUTH, D. (1956). *Brit. J. med. Psychol.*, **29**, 211.

BOYCE, P. R. (1965). *Optica Acta*, **12**, 47.

BUECHNER, H. K., & SCHLOETH, R. (1965). *Zeits. f. Psychol.*, **22**, 207.

BÜHLER, C., & HETZER, H. (1928). *Zeits. f. Psychol.*, **107**, 50.

BURCHARDT, D. (1958). *Mammalia*, **22**, 226.

BURKHOLDER, B. L. (1959). *J. Wildlife Manag.*, **23**, 1.

BURT, W. H. (1943). *J. Mammal.*, **24**, 346.

CALDERWOOD, W. L. (1927). *Salm. Trout Mag.*, **48**, 241.

CAMPBELL, N. R. (1919). *Physics, The Elements*. Cambridge Univ. Press.

CARNAP, R. (1932–3). Psychology in Physical Language, *Erkenntniss*, **111**. Reprinted in Ayer, A. J. (Ed.), *Logical Positivism*. Free Press, Glencoe, 1959.

CARPENTER, C. R. (1958). In Roe, A., & Simpson, G. G., *Behaviour and Evolution*. Yale Univ. Press.

——(1965). In De Vore, L. (Ed.), *Primate Behaviour*. Holt, Rinehart & Winston, N.Y.

CARR, A., & HIRTH, H. (1961). *Anim. Behav.*, **9**, 68.

CARTHY, J. D. (1956). *Animal Navigation*. Allen & Unwin, London.

CASPARI, E. W. (1962). In Washburn, S. L. (Ed.), *Social Life of Early Man*. Methuen, London.

CHAPMAN, F. M. (1935). *Bull. Am. Mus. Nat. Hist.*, **68**, 471.

CHISHOLM, A. H. (1935). *Bird Wonders of Australia*. Angus & Robertson, Sydney.

CLARK, R. S., HERON, W., FEATHERSTONEHAUGH, M. L., FORGAYS, D. H., & HEBB, D. O. (1951). *Canad. J. Psychol.*, **5**, 150.

COLLIAS, N. E., & COLLIAS, E. C. (1956). *Auk*, **73**, 378.

COLLIAS, N. E., & JOOS, M. (1953). *Behaviour*, **5**, 175.

COLLINGWOOD, R. G. (1936). *Proc. Brit. Acad.* XXII.

——(1937). *Proc. Arist. Soc.*, **38**, 85.

COULSON, J. C. (1968). *Nature*, **217**, no. 4127, 478.

COWAN, I. McT. (1947). *Canadian J. Research*, **25**, 139.

CRANE, J. (1941). *Zoologica*, **26**, 145.

CROCE, B. (1941). *History as the Story of Liberty* (trans.). Allen & Unwin, London.

CROOK, J. H. (1964). *Behaviour*, Supp. 10.

——(1965). *Symp. Zool. Soc. Lond.*, **14**, 181.

——(1966). *Symp. Zool. Soc. Lond.*, **18**, 237.

CROOK, J. H., & GARTLAN, J. S. (1966). *Nature*, **210**, 1200.

D'ANCONA, U. (1960). *Symp. Zool. Soc. Lond.*, **1**, 61.
DARLING, F. F. (1946). *A Herd of Red Deer*. Oxford Univ. Press.
DAVIDSON, F. A. (1934). *Bull. Bur. Fish. Wash.*, **48**, 27.
DAVIS, D. E. (1958) *Brit. J. Anim. Behav.*, **6**, 207.
DE VORE, I. (1963). In Rheingold, H. L. (Ed.), *Maternal Behaviour in Mammals*. Wiley, N.Y.
DE VORE, I., & WASHBURN, S. L. (1963). In Howell, F. C., & Bourlière, F. (Eds.), *African Ecology and Human Evolution*. Methuen, London.
DE VOS, A. (1950). *J. Mammal.*, **31**, 169.
DICE, L. R. (1952). *Natural Communities*. Univ. Michigan Press.
DILTHEY, W. (1894). Ideen über eine beschreibende und zergliedernde Psychologie. *Gessammelte Schriften*, Band V., B. G. Teubner, Leipzig and Berlin, 1924.
DREHER, J. J., & EVANS, W. E. (1964). In Tavolga, W. N. (Ed.), *Marine Acoustics (A Symposium)*. Pergamon Press, London.
DURFEE, H., and WOLF, K. (1933). *Zeits. f. Kinderforsch.*, **42**, 273.

EIMERL, S., & DE VORE, I. (1966). *The Primates*. Time-Life Int. (Nederland) N.V.
ESPMARK, Y. (1964). *Anim. Behav.*, **12**, 420.
ESTES, R. D. (1963). *Nat. Geogr. Soc.* Cornell Univ. Sec. Quart. Report (April–June).
——(1969). *Zeits. f. Tierpsychol.*, **26**, 284.

FABRICIUS, E. (1951). *Proc. Xth Int. Ornithol. Cong.*
FOERSTER, R. E. (1934). *Contr. Can. Biol. Fish*, **8**, 347.
——(1936). *J. biol. Bd. Can.*, **2**, 311.
——(1937). *J. biol. Bd. Can.*, **3**, 26.
FRISCH, K. VON (1967). *The Dance Language and Orientation of Bees* (trans.). Oxford Univ. Press.
FRISCH, O. VON (1957). *Zeits. f. Tierpsychol.*, **14**, 231.
FUKUHARA, F. M. (1955). *Comml. Fish. Rev.*, **17**, (3), 1.

GARNER, R. L. (1892). *The Speech of Monkeys*. Heinemann, London.
GEE, E. P. (1953). *J. Bombay Nat. Hist. Soc.*, **51**, 765.
GEIST, V. (1963). *Behaviour*, **20**, 377.
——(1966) *Behaviour*, **27**, 175.
——(1968). *Zeits. f. Tierpsychol.*, **25**, 199.
GLUECK, S., & GLUECK, E. T. (1934). *One Thousand Juvenile Delinquents*. Harvard Univ. Press, Cambridge, Mass.
GOLDFARB, W. (1943a). *J. exp. Educ.*, **12**, 106.
——(1943b). *Am. J. Orthopsychiat.*, **13**, 249.
——(1944a). *Am. J. Orthopsychiat.*, **14**, 162.
——(1944b). *Am. J. Orthopsychiat.*, **14**, 441.
——(1945). *Am. J. Psychiat.*, **102**, 18.
GOLDSMITH, T. H., & GRIFFIN, D. R. (1956). *Biol. Bull.*, **111**, 235.
GOTTLIEB, G. (1965). *Science*, **147**, 1596.
——(1966). *Anim. Behav.*, **14**, 282.

GOTTLIEB, G. (1967). *Abstr. Xth. Int. Conf. Ethology*, Stockholm.
GOULD, E. (1955). *J. Mammal.*, **36**, 399.
GRABOWSKI, U. (1941). *Zeits. f. Tierpsychol.*, **4**, 326.
GRIFFIN, D. R. (1953). *Proc. Nat. Acad. Sci.*, **39**, 884.
——(1958). *Listening in the Dark*. Yale Univ. Press.
——(1960). *Echoes of Bats and Men*. Heinemann, London.
GRIFFIN, D. R., & GOLDSMITH, T. H. (1955). *Biol. Bull.*, **108**, 264.
GRIFFIN, R., WEBSTER, F. A., & MICHAEL, C. R. (1960). *Anim. Behav.*, **8**, 141.
GROOT, C. (1965). *Behaviour*, Supp. 14, 198.
GRUBB, P., & JEWELL, P. A. (1966). *Symp. Zool. Soc. Lond.*, **18**, 179.
GUDGER, E. W. (1949). *Zoologica*, **34**, 10, 99.

HAFEZ, E. S. E. (Ed.) (1962). *The Behaviour of Domestic Animals*. Baillière, Tindall & Cox, London.
HALL, K. R. L. (1963). *Symp. Zool. Soc. London*, **10**, 1.
HALL, K. R. L., & DE VORE, I. (1965). In De Vore, I. (Ed.), *Primate Behaviour*. Holt, Rinehart and Winston, N.Y.
HAMBURG, D. A. (1962). In Washburn, S. L. (Ed.), *Social Life of Early Man*. Methuen, London.
HAMILTON, W. A. (1933). *J. Mammal.*, **14**, 155.
HARDEN JONES, F. R. (1968). *Fish Migration*. Arnold, London.
HARLOW, H. F., & ZIMMERMAN, R. R. (1959). *Science*, **30**, 421.
HARLOW, H. F., & HARLOW, M. K. (1961). *Natural Hist. LXX (10)*, 48.
HASLER, A. D. (1960). *Ergebnisse der Biologie*, **23**, 94.
HASLER, A. D., HORRAL, R. M., WISBY, W. J., & BRAEMER, W. (1958). *Limnol. Oceanography*, **3**, 353.
HASLER, A. D., & WISBY, W. J. (1951). *Am. Nat.*, **85**, 223.
HEDIGER, H. (1948). *Behaviour*, **1**, 130.
——(1949). *Bijd Dierk*, **28**, 172.
——(1962). In Washburn, S. L. (Ed.), *Social Life of Early Man*. Methuen, London.
HEINROTH, O. (1911). *Verhandlungen des V. Int. Orn. Kongress, Berlin*, 587.
HERRICK, C. J. (1956). *The Evolution of Human Nature*. Univ. of Texas Press.
HERSHER, L., RICHMOND, J. B., & MOORE, A. U. (1963). *Behaviour*, **20**, 311.
HESS, E. H. (1959). *Science*, **130**, 133.
HESS, W. R. (1967). *Psychologie in Biologische Sicht*. Thieme, Stuttgart, 2nd edn.
HESS, W. R., & BRÜGGER, M. (1943). *Helv. Physiol. Pharmacol. Acta*, **1**, 33.
HINDE, R. A., THORPE, W. H., & VINCE, M. A. (1956). *Behaviour*, **9**, 214.
HOFFMAN, K. (1954). *Zeits. f. Tierpsychol.*, **11**, 453.
HOGAN-WARBURG, A. J. (1966). *Ardea*, **54**, 111.
HOLST, E. VON, & ST PAUL, U. VON (1960). *Anim. Behav.*, **11**, 1.

HOWELL, F. C., & BOURLIÈRE, F. (Eds.) (1963). *African Ecology and Human Evolution.* Methuen, London.
HULL, C. L. (1943). *Principles of Behaviour.* Appleton Century Crofts, New York.
——(1951). *Essentials of Behaviour.* Yale Univ. Press.
——(1952). *A Behaviour System.* Yale Univ. Press.
HUME, D. (1898). *A Treatise of Human Nature.* Ed. Green, T. H., & Grose, T. H. Longmans, London.
HUMPHREY, M., & OUNSTED, C. (1963). *Brit. J. Psychiat.*, **109**, 599.
HUNTSMAN, A. G. (1942). *Science,* **95**, 381.
HUXLEY, J. S. (1914). *Proc. Zool. Soc. Lond.,* **25**, 491.

IMANISHI, K. (1957). *Psychologia,* **1**, 47.
IMMELMANN, K. (1967). *Abstr. Xth Int. Conf. Ethology,* Stockholm.
ITANI, J. (1963). In Southwick, C. H., *Primate Social Behaviour.* Van Nostrand, N.Y.

JAMES, H. (1959). *Canad. J. Psychol.,* **13**, 59.
JAY, P. (1963). In Southwick, C. H., *Primate Social Behaviour.* Van Nostrand, N.Y.
——(1963). In Rheingold, H. L., *Maternal Behaviour in Mammals.* Wiley, N.Y.
——(1965). In De Vore, I. (Ed.), *Primate Behaviour.* Holt, Rinehart & Winston, N.Y.
JENKINS, D. (1961). *Ibis,* **103a**, no. 2, 155.
JENKINS, D., WATSON, A., & MILLER, G. R. (1963). *J. Anim. Ecol.,* **32**, 317.
JENNINGS, H. S. (1906). *The Behaviour of Lower Organisms.* Columbia Univ. Press, N.Y.
JEWELL, P. A. (1966). *Symp. Zool. Soc. Lond.,* **18**, 85.
JOHNSON, W. E., & GROOT, C. (1963). *J. Fish. Res. Bd. Can.,* **20**, 919.

KAILA, E. (1932). *Ann. Univ. Aboensis,* **17**, 1.
KATZ, I. (1949). *Zoologica,* **34**, 3, 9.
KAWAMURA, S. (1963). In Southwick, C. H., *Primate Social Behaviour.* Van Nostrand, N.Y.
KELLOG, W. N. (1958). *Science,* **128**, 982.
——(1961). *Porpoises and Sonar.* Chicago Univ. Press.
KELSON, H. (1941). *Philos. of Science,* **8**, 553.
——(1943). *Society and Nature.* Univ. Chicago Press.
KELVIN, T. W. (1884). *Lectures on Molecular Dynamics and the Wave Theory of Light.* Baltimore.
KENYON, K. W., & RICE, D. W. (1958). *Condor,* **60**, 3.
KILEY-WORTHINGTON, M. (1965). *Mammalia,* **29**, 177.
KLEBER, E. (1935). *Z. vergl. Physiol.,* **22**, 221.
KLINGHAMMER, E., & HESS, E. H. (1964). *Science,* **146**, 265.
KLOPFER, P. H. (1959a). *Wilson Bull.,* **71**, 262.
——(1959b). *Ecology,* **40**, 90.

KLOPFER, P. H., ADAMS, D. K., & KLOPFER, M. S. (1964). *Proc. Nat. Acad. Sciences*, **52**, 911.

KLOPFER, P. H., & GAMBLE, J. (1966). *Zeits. f. Tierpsychol.*, **23**, 588.

KLOPFER, P. H., & HAILMAN, J. P.(1964). *Zeits. f. Tierpsychol.*, **21**, 755.

KLOPFER, P. H., & KLOPFER, M. S. (1968). *Zeits. f. Tierpsychol.*, **25**, 862.

KOFORD, C. B. (1957). *Ecological Monographs*, **27**, 153.

——(1963). In Southwick, C. H. (Ed.), *Primate Social Behaviour*. Van Nostrand, N.Y.

KORTLANDT, A. (1962). *Scient. American*, **206** (5), 128.

KORTLANDT, A., & KOOIJ, M. (1963). *Symp. Zool. Soc. Lond.*, **10**, 61.

KORTLANDT, A., & VAN ZON, J. C. J. (1967). *Proc. Xth Int. Ethol. Conf.*, Stockholm.

KRAMER, G. (1949). In *Ornithologie als biologische Wissenschaft.* Heidelberg.

——(1951). *Proc. Xth Int. Cong. Orn.* Uppsala, 271.

——(1952). *Ibis*, **94**, 265.

KRAMER, G., & REISE, E. (1952). *Zeits. f. Tierpsychol.*, **9**, 620.

KRUIJT, J. P. (1962). *Symp. Zool. Soc. Lond.*, **8**, 219.

——(1964). *Behaviour*, Supp. XII.

KRUIJT, J. P., & HOGAN, J. A. (1967). *Ardea*, **55**, 203.

LACK, D. (1953). *The Life of the Robin*. Pelican Books, London.

——(1954). *The Natural Regulation of Animal Numbers*. Oxford Univ. Press.

——(1968). *Ecological Adaptations for Breeding in Birds*. Methuen, London.

LAMB, F. B. (1954). *Natural History*, **53**, 231.

LAWICK-GOODALL, J. VAN (1968). *Anim. Behav. Monog.*, vol. 1, pt. 3.

LEAGUE OF NATIONS. *Prostitutes: Their Early Lives*. Geneva, 1938.

LEIBOWITZ, H. W. (1955). *J. Exp. Psychol.*, **49**, 209.

LENT, P. C. (1966). *Zeits. f. Tierpsychol.*, **23**, 701.

LEUTHOLD, W. (1966). *Behaviour*, **27**, 213.

LEVY, D. M. (1944). *Am. J. Orthopsychiat.*, **14**, 644.

LEYHAUSEN, P. (1965). *Symp. Zool. Soc. Lond.*, **14**, 249.

LEYHAUSEN, P., & WOLFF, R. (1959). *Zeits. f. Tierpsychol.*, **16**, 666.

LIDDELL, H. (1960). In Tanner, J. M. (Ed.), *Stress and Psychiatric Disorder. Experimental Neurosis in Animals*. Blackwell, Oxford.

LINDAUER, M. (1957). *Naturwissenschaften*, **44**, 1.

——(1959). *Z. Vergl. Physiol.*, **42**, 43.

——(1960). *Cold Spring Harbour Symposium on Quantitative Biology*, **25**, 371.

LOCKLEY, R. M. (1942). *Shearwaters*. Dent, London.

LORENZ, K. Z. (1935). *J. Orn.*, **83**, 137, 289.

——(1937). *Auk*, **54**, 245.

——(1941). *J. Ornithol.*, **89** (Suppl.), 194.

——(1950). *Soc. Exp. Biol. Symp. IV*, 221.

——(1952). *King Solomon's Ring*. Methuen, London.

LORENZ, K. Z. (1955). In Schaffner, B. (Ed.), *Group Processes*. Jos. Macey Foundation, N.Y.
——(1957). In Schiller, C. H. (Ed.), *The Nature of Instinct*. International Univ. Press, N.Y.
——(1958). Reprint from *Scientific American* in *Psychobiology*. W. H. Freeman, San Fran., 1966.
——(1960). *Fort. der Zool.*, **12**, 263.
——(1966). *On Aggression*. Methuen, London.
LOWE, V. P. W. (1966). *Symp. Zool. Soc. Lond.*, **18**, 211.

MCBRIDE, G. (1967). *Abstr. Xth. Int. Conf. Ethol.*, Stockholm.
——(1968). *Proc. Symp. Impact of Civilisation on the Biology of Man*. Austral. Acad. Sci.
MCBRIDE, G., PARER, I. P., & FOENANDER, F. (1969). *Anim. Behav. Monographs*, vol. 2, pt. 3, 127.
MCDOUGALL, Wm (1936). *An Introduction to Social Psychology*, 23rd ed. Methuen, London.
MAKKINK, G. F. (1936). *Ardea*, **25**, 1.
MALCOLM, N. (1964). In Wann, T. W. (Ed.), *Behaviourism and Phenomenology*. Chicago Univ. Press.
MALINOWSKI, B. (1923). *Psyche*, **4**, 99.
MARLER, P. (1956). *Ibis*, **98**, 231.
——(1965). In De Vore, I. (Ed.), *Primate Behaviour*. Holt, Rinehart & Winston, N.Y.
MARSHALL, A. J. (Ed.) (1960). *Biology and Comparative Physiology of Birds*, vol. 1. Academic Press, New York and London.
MAST, S. O. (1932). *Physiological Zoology*, *V*, 1, 1.
MAST, S. O., & HAHNERT, W. F. (1935). *Physiological Zoology*, *VIII*, 3, 255.
MAST, S. O., & PUSCH, L. C. (1924). *Biological Bulletin*, **46**, 2, 55.
MATTHEWS, G. V. T. (1953). *J. Exp. Biol.*, **30**, 370.
——(1961). *Ibis*, **103a**, 211.
——(1963). *Proc. XIIIth Int. Ornithol. Cong.*, Ithaca, 415.
——(1968). *Bird Navigation*. Cambridge Univ. Press.
MATTHEWS, L. H. (1939). *Proc. Zool. Soc. Lond.*, Series A, 109, 43.
MATURANA, H. R., & FRENK, S. (1963). *Science*, **142**, 977.
——(1965). *Science*, **150**, 359.
MEEHL, P. E. (1954). *Clinical Versus Statistical Prediction*. Univ. Minnesota Press.
MELZACK, R., & THOMPSON, W. R. (1956). *Canad. J. Psychol.*, **10**, 82.
MENNER, E. (1938). *Zool. Jb., Abt. allg. Zool. Physiol.*, **48**, 481.
MENUT, G. (1943). *La dissociation familiale et les troubles du caractère chez l'enfant*. Paris.
MILLER, G. R., JENKINS, S. D., & WATSON, A. (1966). *J. Appl. Ecol.*, **3**, 313.
MURIE, A. (1944). *U.S. Dept. Int. Natl. Park Service Fauna Series*, 238.

NEWTON, I. (1931). *Opticks*, 4th ed. Bell, London.
——(1934). *Principia* (trans. by A. Motte, revised by A. Cajori). Univ. California Press.
NICE, M. (1953). *The Condor*, **55**, 33.
NISSEN, H. W., CHOW, K. L., & SEMMES, J. (1951). *Am. J. Psychol.*, **64**, 485.
NOBLE, G. K., & CURTIS, B. (1939). *Bull. Am. Mus. Nat. Hist.*, **76**, 1.
NUBLING, E. (1939). *Emu*, **39**, 22.
NUBLING, G. E. (1941). *The Australian Zoologist*, **10**, 1, 95.

O'FARRELL, T. P. (1965). *J. Mammal.*, **46**, 406.

PALMER, R. S. (1941). *Proc. Boston Soc. Nat. Hist.*, **42**, 1.
PARDI, L., & PAPI, F. (1952). *Naturwissenschaften*, **39**, 262.
——(1953). *Z. vergl. Physiol.*, **35**, 459.
PAYNE, R. S. (1969). *Proc. XIth Int. Ethol. Cong.*, Rennes.
PEARSE, A. S. (1913). *Ann. Rep. Smithson Inst. Wash.*, 415.
PENNEY, R. L., & EMLEN, J. T. (1967). *Ibis*, **109**, 99.
PERDECK, A. C. (1958). *Ardea*, **46**, 1.
PFEFFER, P. (1967). *Mammalia*, **31** (Supp.), 1.
PIAGET, J. (1926a). *The Child's Conception of the World* (trans.). Kegan Paul, London.
——(1926b). *The Language and Thought of the Child* (trans.). Kegan Paul, London.
PILLERI, G., & KNUCKEY, J. (1969). *Zeits. f. Tierpsychol.*, **26**, 48.
PITMAN, C. R. S. (1931). *A Game Warden Among His Charges*. Nisbet, London.
PITZ, G. F., & ROSS, R. B. (1961). *J. Comp. physiol. Psychol.*, **54**, 602.
PORTMAN, A. (1961). *Animals as Social Beings*. Viking Press, N.Y.
PRATT, A. (1951). *The Lore of the Lyrebird*. Robertson & Mullens, Melbourne.
PRITCHARD, A. L. (1939). *J. Fish. Res. Bd. Can.*, **6**, 217.
——(1944a). *J. Fish. Res. Bd. Can.*, **6**, 217.
——(1944b). *Copeia*, **80**, 2.
PRUGH, D. G., STAUB, E. M., & SANDS, H. H. (1953). *Am. J. Orthopsychiatr.*, **23**, 70.
PRUITT, W. O. (Jr.) (1960). *Biol. Papers Univ. Alaska*, no. 3, 1.
——(1965). *J. Mammal.*, **46**, no. 2, 350.
PUMPHREY, R. J. (1948). *Ibis*, **90**, 171.
——(1961). Sensory Organs: Vision. In Marshall, A. J. (Ed.), *Biology and Comparative Physiology of Birds*. Academic Press, New York and London.

REISEN, A. H. (1966). In Stellar, E., and Sprague, J. M. (Eds.), *Progress in Physiological Psychology*. Academic Press, New York & London.
REYNOLDS, V., & REYNOLDS, F. (1965). In De Vore, I. (Ed.), *Primate Behaviour*. Holt, Rinehart & Winston, N.Y.
RIBBLE, M. A. (1943). *The Rights of Infants*. Columbia Univ. Press, N.Y.

RICH, W. H., & HOLMES, H. B. (1929). *Bull. Bur. Fish. Wash.*, **44**, 215.
ROE, A., & SIMPSON, G. G. (1958). *Behaviour and Evolution.* Yale Univ.
    Press.
ROEDER, K., & TREAT, A. (1957). *J. Exp. Zool.*, **134**, 127.
ROEDER, K. D., & TREAT, A. (1960). *Proc. XI Internat. Ent. Congr.*
——(1961). *Am. Scient.*, **49**, 135.
ROWAN, W. (1946). *Trans. Roy. Soc. Canada*, **40**, 123.
——(1950). *J. Mammal.*, **31**, 2, 167.
RUSSELL, E. S. (1946). *The Behaviour of Animals.* Arnold, London.

SALZEN, E. A. (1967). *Behaviour*, **28**, 232.
SAUER, E. G. F. (1957). *Zeits. f. Tierpsychol.*, **14**, 29.
SAUER, E. G. F., & SAUER, E. M. (1955). *Rev. Suisse Zool.*, **62**, 250.
SAUNDERS, R. L., KERSWILL, C. J., & ELSON, P. F. (1965). *J. Fish. Res.
    Bd. Can.*, **22**, 625.
SCHAFFER, H. R. (1958). *Brit. J. Med. Psychol.*, **31**, 174.
SCHAFFER, H. R., & EMERSON, P. (1964). *Monog. of Soc. for Research
    in Child Development*, **94**, vol. 29, no. 3.
SCHALLER, G. B. (1963). *The Mountain Gorilla: Ecology and Behaviour.*
    Univ. Chicago Press.
SCHALLER, G. B., & EMLEN, J. T. (1963). In Howell, F. C., & Bourlière,
    F. (Eds.), *African Ecology and Human Evolution.* Methuen, London.
SCHENKEL, R. (1948). *Behaviour*, **1**, 81.
——(1966a). *Symp. Zool. Soc. Lond.*, **18**, 11.
——(1966b). *Z. Saugetierk*, **31**, 177.
SCHEVILL, W. E., & LAWRENCE, B. (1949). *Science*, **109**, 143.
——(1953). *J. Exp. Zool.*, **124**, 147.
SCHJELDERUP-EBBE, T. (1922). *Zeits. f. Psychol.*, **88**, 225.
SCHLOETH, R. (1961). *Zeits. f. Tierpsychol.*, **18**, 574.
SCHMIDT, J. (1906). *Rapp. P.-v. Réun. Cons. perm. int. Explor. Mer*, **5**,
    137.
——(1922). *Phil. Trans. R. Soc. B.*, **211**, 179.
——(1932). *Danish Eel Investigations During Twenty Years (1905–1930).*
    Carlsberg Foundation.
SCHMIDT-KOENIG, K. (1958). *Zeits. f. Tierpsychol.*, **15**, 301.
SCHUTZ, F. (1965). *Psychologische Forschung*, **28**, 439.
——(1966). *Stud. Gen.*, **19**, 273.
SCHÜZ, E. (1938). *Vogelzug*, **9**, 65.
——(1949). *Vogelwarte*, **15**, 63.
——(1950). *Bonner Zool. Beitr.*, **1**, 239.
SETON, E. T. (1910). *Life Histories of Northern Mammals*, 2 vols.
    Constable, London.
SHAPALOV, L. (1937). *Calif. Fish Game*, **23**, 205.
SIMKISS, K. (1963). *Bird Flight.* Hutchinson, London.
SIMMONS, K. E. L. (1954–5). *Agricultural Magazine*, **60–1**, 3, 93, 131,
    181, 235, 294.
——(1959). In Banister, D. A. (Ed.), *The Birds of the British Isles*, vol.
    VIII. Oliver & Boyd, Edinburgh.

SIMMONS, K. (1970). In Crook, J. H., *Social Behaviour in Birds and Mammals*. Academic Press, New York and London.

SKINNER, B. F. (1953). *Science and Behaviour*, Free Press, N.Y., rpt., 1966.

SLADEN, W. J. L. (1958). *Falkland Islands Dependencies Survey*, no. 17.

SMITH, F. V. (1960). *Anim. Behav.*, **8**, 197.

——(1962). *Symp. Zool. Soc. Lond.*, **8**, 171.

——(1965). *Anim. Behav.*, **13**, 84.

SMITH, F. V., & BIRD, M. W. (1963a). *Anim. Behav.*, **11**, 57.

——(1963b). *Anim. Behav.*, **11**, 300.

——(1963c). *Anim. Behav.*, **11**, 397.

——(1964a). *Anim. Behav.*, **12**, 60.

——(1964b). *Anim. Behav.*, **12**, 252.

——(1964c). *Anim. Behav.*, **12**, 259.

SMITH, F. V., & HARDING, L. W. (unpublished). The effect of a conventional reinforcer on the imprinting-type response.

SMITH, F. V., & HOYES, P. A. (1961). *Anim. Behav.*, **9**, 159.

SMITH, F. V., & NOTT, K. (1970). *Zeits. f. Tierpsychol.*, **27**, 108.

SMITH, F. V., & TEMPLETON, W. B. (1966). *Anim. Behav.*, **14**, 291.

SMITH, F. V., VAN-TOLLER, C., & BOYES, T. (1966). *Anim. Behav.*, **14**, 120.

SOUTHWICK, C. H. (Ed.) (1963). *Primate Social Behaviour*. Van Nostrand, N.Y.

SOUTHWICK, C. H., MIRZA, A. B., & SIDDIQI, M. R. (1965). In De Vore, I. (Ed.), *Primate Behaviour*. Holt, Rinehart & Winston, N.Y.

SPALDING, D. (1873). Rpt. in *Brit. J. Anim. Behav.*, 1954, **2**, 2.

SPITZ, R. A. (1946). *Genet. Psychol. Monog.*, **34**, 57.

——(1946). *The Psychoanalytic Study of the Child*, **2**, 313.

SPITZ, R., & WOLF, K. (1945). *The Psychoanalytic Study of the Child*, **1**, 53.

STORR, A. (1968). *Human Aggression*. Allen Lane, Penguin Press, London.

STOTT, D. H. (1950). *Delinquency and Human Nature*. Carnegie U.K. Trust, Dunfermline.

STRUHSAKER, T. (1967). *Behaviour*, **29**, 81.

TAFT, A. C., & SHAPALOV, L. (1938). *Calif. Fish Game*, **24**, 118.

TALBOT, L. M., & TALBOT, M. H. (1963). *Wildlife Monog.*, **12**, 1.

TAVOLGA, W. N. (Ed.) (1964). *Marine Bio-Acoustics*. Pergamon Press, Oxford.

THOMPSON, D. Q. (1952). *J. Mammal.*, vol. 33, no. 4, 429.

THOMPSON, W. R., & HERON, W. (1954). *Canad. J. Psychol.*, **8**, 17.

THORPE, W. H. (1963). *Learning and Instinct in Animals*. Methuen, London, 2nd ed.

THORPE, W. H., & WILKINSON, D. H. (1946). *Nature*, **158**, 903.

TINBERGEN, N. (1951). *The Study of Instinct*. Clarendon Press, Oxford.

——(1953). *The Herring Gull's World*. Collins, London.

TOLMAN, E. C. (1932). *Purposive Behaviour in Animals and Men*. Appleton Century Co., N.Y.

TROYER, W. A., & HENSEL, R. J. (1964). *J. Wildl. Mgmt.*, **28**, 769.
TUCKER, D. W. (1959). *Nature*, **183**, 495.

VAUGHT, R. W. (1964). *J. Wildl. Mgmt.*, **28**, 208.
VERHEYEN, R. (1954). *Inst. Parcs Nat. Congo Belge, Brussels.* Monographie éthologique de l'hippopotame (*Hippopotamus amphibius*).
VERWEY, J. (1930). *Treubia*, **12**, 167.
VINCE, M. A. (1960). *Behaviour*, **15**, 219.
——(1964). *Anim. Behav.*, **12**, 531.
——(1966a). *Anim. Behav.*, **14**, 34.
——(1966b). *Anim. Behav.*, **14**, 389.

WAHL, O. (1932). *Z. vergl. Physiol.*, **16**, 529.
——(1933). *Z. vergl. Physiol.*, **18**, 709.
WALTHER, F. R. (1958). *Zeits. f. Tierpsychol.*, **15**, 340.
——(1960). *Zeits. f. Tierpsychol.*, **17**, 188.
——(1964a). *Zeits. f. Tierpsychol.*, **21**, 393.
——(1964b). *Zeits. f. Tierpsychol.*, **21**, 871.
——(1965). *Zeits. f. Tierpsychol.*, **22**, 167.
——(1966). *Z. Saugetierk.*, **31**, 1.
WASHBURN, S. L., & AVIS, V. (1958). In Rose, A., & Simpson, G. G. (Eds.), *Behaviour and Evolution.* Yale Univ. Press.
WATT, G. (1951). *The Farne Islands: Their History and Wild Life.* Country Life, London.
WEIDMANN, U. (1958). *Zeits. f. Tierpsychol.*, **15**, 277.
WERNER, H. (1940). *The Comparative Psychology of Mental Development.* Harper, N.Y.
WHITMAN, C. O. (1898). *Anim. Behaviour*, Biol. Lect. Marine Biol. Lab., Wood's Hole, Mass., 285–338.
WINN, H. E., SALMON, M., & ROBERTS, N. (1964). *Zeits. f. Tierpsychol.*, **21**, 798.
WISBY, W. J., & HASLER, A. S. (1954). *J. Fish. Res. Bd. Can.*, **11**, 472.
WRIGHT, R. W. (1954). *Int. Pacific Salmon Fisheries Commission Ann. Report for 1953.* I.P.F.S.C., 1954.
WYNNE-EDWARDS, V. C. (1962). *Animal Dispersion in Relation to Social Behaviour.* Oliver & Boyd, Edinburgh.

YERKES, R. M. (1912). *J. Anim. Behav.*, **2**, 332.

ZAHAVI, A. (1971). *Ibis*, **113**, 203.

INDEX OF FIRST AUTHORS

GENERAL INDEX